P9-BYW-662

ALSO BY DESMOND MORRIS

The Biology of Art
The Mammals: a Guide to the living Species
Men and Snakes (co-author)
Men and Apes (co-author)
Men and Pandas (co-author)
Zootime
Primate Ethology (editor)
The Naked Ape
The Human Zoo
Patterns of Reproductive Behaviour
Intimate Behaviour
Manwatching: a Field-guide to Human Behaviour
Gestures: Their Origins and Distributions (co-author)
Animal Days (autobiography)
The Soccer Tribe
The Giant Panda (co-author)
Inrock (fiction)
The Book of Ages
The Art of Ancient Cyprus
Bodywatching: a Field-guide to the Human Species
Catwatching
Dogwatching
The Secret Surrealist
Catlore
The Human Nestbuilders
Horsewatching
The Animal Contract
Animalwatching: a Field-guide to Animal Behaviour
Babywatching
Christmas Watching
The World of Animals
The Naked Ape Trilogy
The Human Animal: a Personal View of the Human Species
Bodytalk: A World Guide to Gestures
Catworld: a Feline Encyclopedia
The Human Sexes: A Natural History of Man and Woman
Cool Cats: the 100 Cat Breeds of the World
Body Guards: Protective Amulets and Charms
The Naked Ape and Cosmetic Behaviour (co-author) (in Japanese)
The Naked Eye (autobiography)
Dogs: a Dictionary of Dog Breeds
Peoplewatching
The Silent Language (in Italian)
The Nature of Happiness (in Italian)

THE NAKED WOMAN

THE NAKED WOMAN

A Study of the Female Body

Desmond Morris

THOMAS DUNNE BOOKS
ST. MARTIN'S PRESS ✹ NEW YORK

THOMAS DUNNE BOOKS.
An imprint of St. Martin's Press.

THE NAKED WOMAN. Copyright © 2004 by Desmond Morris. All rights
reserved. Printed in the United States of America. No part of this book
may be used or reproduced in any manner whatsoever without written
permission except in the case of brief quotations embodied in critical arti-
cles or reviews. For information, address St. Martin's Press, 175 Fifth
Avenue, New York, N.Y. 10010.

www.stmartins.com

ISBN 0-312-33852-X
EAN 978-0-312-33852-7

First published in Great Britain by Jonathan Cape

10 9 8 7 6 5 4 3 2

CONTENTS

PICTURE CREDITS

ACKNOWLEDGEMENTS

I would like to express my special thanks to the following: my wife Ramona for her endless encouragement and constructive criticism; my colleague Clive Bromhall, for many valuable discussions; and from Random House, Marcella Edwards, Caroline Michel, Dan Franklin and Ellah Allfrey, for their editorial expertise, Nadine Bazar for her painstaking picture research and David Fordham for his inset design.

INTRODUCTION

This book takes the reader on a guided tour of the female body, explaining its many unusual features. It is not a medical text, or a psychologist's laboratory analysis, but a zoologist's portrait, celebrating women as they appear in the real world, in their natural environment.

The human female has undergone dramatic changes during the course of her evolution – far more than the human male. She has left behind many of the feminine qualities of other primates and, in the shape of modern woman, has become a unique being of an extraordinary kind.

Every woman has a beautiful body – beautiful because it is the brilliant end-point of millions of years of evolution. It is loaded with amazing adjustments and subtle refinements that make it the most remarkable organism on the planet. Despite this, at different times and in different places, human societies have tried to improve on nature, modifying and embellishing the female body in a thousand different ways. Some of these cultural elaborations have been pleasurable, others have been painful, but all have sought to make the human female even more beautiful than she already is.

Local concepts of beauty have varied wildly and each human society has developed its own ideas of what is more appealing. Some cultures like slender figures, others prefer more rounded flesh; some like small breasts, others relish large ones; some like white teeth, others insist on filed teeth; some like shaven heads, others dote on long, luxuriant hair. Even within Western culture there have been striking contrasts as the fickle world of fashion keeps on changing its priorities.

As a result, each chapter – as the book travels from head to toe – not only explains the exciting biological features that all human females share, but also discusses the many ways in which these features have been exaggerated or suppressed, enlarged or reduced, and in this way attempts to give a rounded picture of the most fascinating subject in the world – the naked woman.

On a personal note – this book reflects a lifelong fascination with the evolution and status of the human female. A few years ago this led me to make a series for American television called *The Human Sexes* in which I examined in some detail the nature of the relationship between human males and females, all around the globe. The more I travelled, the more disturbed and angry I became with the way women were being treated in many countries. Despite the advances made by feminist rebellion in the West, there are still millions of women in other parts of the world who are considered the 'property' of males and as inferior members of society. For them, the feminist movement simply did not happen.

To me, as a zoologist who has studied human evolution, this trend towards male domination is simply not in keeping with the way in which *Homo sapiens* has developed over a period of millions of years. Our success as a species was due to a division of labour between males and females, in which the males became specialized as hunters. Living in small tribes, this meant that, with the males away hunting, the females were left in the very centre of social life, gathering the food and preparing it, rearing the young, and generally organizing the tribal settlement. As men became better at focusing on their one, crucially important task, women became better at dealing with several problems at once. (This personality difference is still with us today.) There was never any question of one sex being dominant over the other. They relied totally on one another for survival. There was a primeval balance between the human sexes – they were different but equal.

This balance was lost when human populations grew, towns and cities were built, and tribespeople became citizens. Religion, at the centre of human societies, has had a major part to play. In ancient times the great deity was always a woman, but then, as urbanization spread, She underwent a disastrous sex change and, in simple terms, the benign Mother Goddess became the authoritarian God the Father. With a vengeful male God to back them up, ruthless holy men through the ages have ensured their own affluent security and the higher social status of men in general, at the expense of women who sank to a low social status that was far from their evolutionary birthright. It was this birthright that the suffragettes and later the feminists sought to regain. It may be imagined that these women were asking for a new social respect, asserting new rights. But in reality they were simply seeking to have their ancient, primeval role returned to them. In the West, they largely succeeded, but elsewhere the subordination of women has continued to thrive.

After completing *The Human Sexes* I had become more and more preoccupied with this issue and, when it was agreed that a new edition of my 1985 book *Bodywatching* should be prepared, I decided that, instead of following the original pattern and dealing with both sexes, I would confine the new book solely to the female body. In *Bodywatching* I had examined the human body from head to toe, taking each part of it in turn. I have kept that arrangement for the present book, taking the reader on an anatomical tour of inspection, from head to toe, or, to be more precise, from hair to feet. Some of the text from the original *Bodywatching* book has been incorporated, but very little. Although it started out as a revision of an old book, *The Naked Woman* has ended up almost entirely as a new work.

In each chapter I have presented the biological aspect of a particular part of the female body – those aspects that all women share – and I have then gone on to examine the various ways in which different societies have modified these biological qualities. It has been an absorbing voyage of discovery and I only wish that, when I was eighteen, I had known all that I know now – as a result of writing this book – about the complexity of the female form.

1. The Evolution

To the zoologist, human beings are tailless apes with very large brains. Their most astonishing feature is just how incredibly successful they have been. While other apes cower in their last retreats, awaiting the arrival of the chainsaw, the 6,000 million humans have infested almost the entire globe, spreading so far and so fast that, like a plague of giant locusts, they have dramatically changed the landscape.

The secret of their success has been their ability to live in larger and larger populations where, even at the highest densities, they are able to adapt to the stresses of life and continue to breed under conditions that any other ape would find intolerable. Combined with this ability is an insatiable curiosity that keeps them ever searching for new challenges.

This magic combination of friendliness and curiosity has been made possible by an evolutionary process called neoteny, which has seen humans retain juvenile characters into adult life. Other animals are playful when they are young, but lose this quality when they mature. Humans remain playful all their lives – they are the Peter Pan species that never grows up. Of course, once they have become adult, they call play by different names; they refer to it as art or research, sport or philosophy, music or poetry, travel or entertainment. Like childhood play, all these activities involve innovation, risk-taking, exploration and creativity. And it is these activities that have made us truly human.

Men and women have not followed this evolutionary trend in quite the same way. Both have gone a long way down the

1

'childlike-adult' path, but they have advanced at slightly different rates with certain features. Men are slightly more childlike in their behaviour, women in their anatomy. For instance:

At the age of thirty, men are 15 times more accident-prone than women. This is because men have retained the risk-taking element of child's play more strongly than women. Although this quality frequently gets men into trouble, it was a valuable asset back in primeval times when, in order to succeed in the hunt, men were forced to take risks. Primeval women were too valuable to risk on the hunt, but the males of the tribe were expendable, so they became the specialized risk-takers. If a few of them died in the process, it did not reduce the breeding abilities of the small tribes, but if a few women died, then the breeding rate was immediately threatened. It is important to remember that, in primeval times, there were so few of us alive on the planet that breeding rates were all-important.

There are more male inventors than female inventors. Risk-taking was not only physical, it was also mental. Innovation always involves risk – trying out something unknown rather than relying on well-tried, trusted traditions. Women had to be cautious. In their primeval role at the very centre of tribal society, with responsibility for almost everything except hunting, they could not afford to make costly mistakes. During the course of evolution they became better at doing several things at once; they became more fluent verbal communicators; their senses of smell, hearing, touch and colour vision were all superior to those of the males; they became better nurturers – more sensitive parents; and they became more resistant to disease – their health as mothers was vitally important.

All of this added up to a difference in male and female brains, in which men retained more 'little boy' features than women did 'little girl' qualities. Men became more imaginative and sometimes perverse. Women became more sensible and caring. These differences suited their roles in society. They complemented one another and the combination spelled success.

Physically the story was rather different. Because of the new division of labour that was evolving, men had to be physically stronger, and more athletic, for the hunt. The average male body contains 28 kg (57 lb) of muscle, the female only 15 kg (33 lb).

2

The typical male body is 30 per cent stronger, 10 per cent heavier, and 7 per cent taller than the typical female body. The female body, being so important for reproduction, had to be better protected against starvation. As a result, the average woman's curvaceous body contains 25 per cent fat, while the stringy male has only 12.5 per cent.

This greater retention of puppy-fat in the female was a strongly infantile characteristic, and with it went a whole host of other juvenile features that served her well. Adult males had been programmed by evolution to be strongly protective of their children. To thrive, the slow-growing human offspring required the assistance of both parents. Paternal responses to the rounded, fat-covered bodies of human babies were so strong that they could be exploited by the adult females. The more physical baby features the females displayed, the more protective responses they could elicit in their mates.

The result of this was that adult women's voices remained higher pitched than men's. Deep male voices operate at 130–145 cycles a second. High female voices operate at 230–255 cycles a second. In other words, women kept childlike voices. Women also retained more juvenile facial features and, most conspicuously, kept their childlike hair pattern. While adult males grew their heavier brows, chins and noses, and their moustaches, beards and hairy chests, women kept their smooth, finer-boned, baby-faces.

So, to sum up – as the human sexes advanced down their evolutionary pathway, towards greater and greater neoteny, the males *behaved* in a more and more childlike way, while showing fewer physical changes, while the females *developed* more and more childlike physical qualities, while showing fewer childlike mental qualities.

It is important to make a point here about the degree of difference between men and women. I have been concentrating on listing the various differences between the sexes, but it is crucial to remember that both human sexes are 100 times more neotenous in every respect than the sexes of other species. The differences between men and women are very real and very interesting, but they remain very slight. I have dwelt on them here only because it is important to establish, at the start, the fact that the human female's body is

3

more advanced – that is, more neotenous – than the male's in many ways. Understanding this will help to clarify many of the features of the female anatomy that we meet as we travel from head to toe. It does not explain everything, because there have, in addition, been many highly specialized evolutionary developments in female anatomy, particularly in sexual and reproductive features, that make a woman's body such a highly evolved and wonderfully refined organism. As we shall see . . .

2. The Hair

There is scarcely a woman alive today who allows her hair to grow as nature intended. If she did, she would end up with a mane that reached down to her knees, or, if she were dark-skinned, with a huge woolly bush that dominated her skull. Just how our remote, primeval ancestors managed to cope with these extravagant hair patterns, before they had invented knives, scissors, combs and other grooming tools, is never discussed by anthropologists, perhaps because they have no answer. Often, when prehistoric people are described in books, the illustrations show, in their imaginative reconstructions, women who have somehow mysteriously paid a visit to the hairdresser before posing. Their hair is always too short. Unless hairdressing, rather than prostitution, is the world's oldest profession, there is something wrong here, and the error conceals one of the great mysteries of female anatomy – namely, why does the human female grow such ridiculously long tresses? In an ancient, tribal world, such an exaggerated, swishing cape of hair would prove to be a serious encumbrance, reminiscent of a peacock's tail. What was the evolutionary advantage of such an excessive development?

Even odder is the fact that, apart from the top of her head, her armpits and her genitals, the typical human female is virtually hairless. It is true that, under a magnifying glass, it is possible to see tiny, stunted hairs all over her skin, but from a distance these are invisible and her skin is functionally naked. This makes her metre-long head hair even more outlandish.

It is not too difficult to trace the human hair pattern back to its origins. When a chimpanzee foetus is about twenty-six weeks old

5

it displays a hair distribution that is very similar to our adult one. The fact that, in humans, this pattern survives into adulthood is yet another example of neoteny. Unlike the apes, who grow a full coat before they are born, we retain the foetal hair pattern all our lives. Men are less advanced than women in this respect, having hairier bodies, with long moustaches and beards, but both sexes remain functionally naked over most of their body surface. Even the hairiest of males would gain no comfort from his chest-hairs on a freezing night, or avoid sunburn in intense heat.

So it would seem that nature has dealt us an extremely odd hair-style, when we are compared with any other species. The foetal explanation may tell us where we acquired our bizarre adult hair pattern, but it does not tell us what survival advantage we gained from keeping it. Inevitably, where there is no obvious explanation, speculative ideas abound.

Proponents of the aquatic theory of human origin have suggested that we lost our body fur as an adaptation to swimming, but retained our head hair to protect the tops of our heads from the rays of the sun. They have also suggested that the metre-long female head hair was useful for infants to cling on to when swimming with their mothers. Critics of the aquatic theory consider this to be farfetched. If mothers were diving for food in the water, they would have been unlikely to allow their infants to accompany them. Also, if our ancestors evolved in a hot African climate, it is likely that their hair pattern was not long and flowing, but much bushier – closer to that seen on modern African heads.

The idea of scalp hair as protective does, however, have some merit, with or without an aquatic location. If primeval humans became daytime hunter/gatherers on the African savannahs they would require a shield from the intense heat of the tropical sun. Thick head hair would provide that, while keeping the rest of the skin naked would dramatically increase cooling by sweating. (Sweat cools five times as efficiently on naked skin as it does on a furry coat.) If other African animals retained their body fur, this was presumably because they were most active at dawn and dusk, when the sun was not blazing down on them. Early humans were typically daytime animals, like other apes and monkeys.

This may explain the typical African hairstyle – a thick bushy covering over the scalp, efficiently insulating the brain from over-heating – but it does not help to clarify the mystery of the long, flowing hair of humans from the cooler regions to the north. Some anthropologists have suggested that the very long head hair helped to keep the bodies of the northern peoples warm in winter – like a natural cape thrown over the shoulders and hanging down the back. As they crouched up at night, the great mane of hair could have acted almost like a blanket against the bitter cold. It may even have given them the idea for making their very first clothing by wrapping animal skins around their bodies. But if this were the case, why did the cold-country humans not re-grow a whole coat of thick fur to protect them? As before, there are serious flaws in the argument.

The most likely explanation is that the bizarre human hair pattern acts as a species flag – a display that set us apart from all our close relatives (relatives that we have long since eliminated). If we try to picture a little group of our remote ancestors, long before they developed clothing or any kind of cutting implement, it is clear that they would look very different from anything else on the planet. With their naked bodies surmounted by long swishing capes or gigantic woolly bushes, they would immediately be identifiable as members of this newfangled species that walks about on its hind legs. This may seem an odd way to label a species, but a quick look at the other apes and monkeys soon shows how often strange hair patterns have arisen as species identification markers. There is a rich variety of crests, manes, capes, beards, moustaches and brightly coloured hair patches. Primates are predominantly visual animals and it follows that displaying conspicuous visual signals will be the quickest and most efficient way of distinguishing one species from another.

In their primeval condition, our remote human ancestors, with their naked bodies and long head hair, could be spotted far off in the distance, and easily differentiated from their furry-bodied cousins. Coming slightly closer, it would then become possible to distinguish between the sexes. The males, with their hairy faces, could not be confused with the naked-faced females.

There is, however, more to the human hair patterns than just species and gender identification. As human beings began to spread out from their original homeland in Africa and were forced to adapt to different environments, these new peoples started to differ more and more from the tropical ones they left behind. The need to adapt to different climates set them off on evolutionary pathways leading to the development of several distinct racial types. Finding themselves struggling to survive in hot, dry deserts, or in moderately warm temperate zones, or in the freezing northlands, their bodies had to become modified if they were to survive. Once these modifications had been achieved it was important they should not be lost. As with any other evolutionary trend, barriers had to be set up that would reduce interbreeding. The different races had to look as different from one another as possible. One of the quickest ways to achieve this was by varying the human hair pattern. Woolly hair, crinkly hair, wavy hair, straight hair, blond hair – variations of this type could quickly label human groups as being different from one another.

This process obviously started to gain momentum at an early stage, as humans spread their range wider and wider across the globe. There is little doubt that we were on the way to evolving as a new group of closely related species – tropical humans, desert humans, temperate humans, polar humans, and so on. Our different hairstyles were the first hint that this process was taking place. But before it could get very far, the human story took a dramatic new turn. Through our advanced intelligence, we became incredibly mobile. We invented boats and ships, we tamed horses and rode them, we invented the wheel and made coaches, we built trains and cars, railways and motorways, and eventually aeroplanes. The racial differences that had started to develop were still at a very preliminary stage. Only two kinds had made any headway – those concerned purely with heat and humidity (differences in skin pigmentation, the density of sweat glands, and such features) and those concerned with visual labels – the hair patterns.

Modern human populations have little use today for the climatic adaptations of their bodies. They are specializations that have become almost obsolete. We have learnt to tame our environments with our clothing, with fire and central heating, with refrigeration and air conditioning. The surviving differences between the races are no longer important. As for the different hair patterns that arose as isolating mechanisms, helping to keep the different types apart, they are today nothing but an outdated nuisance. As we no longer do keep apart, but mix together all over the world, they only lead to disharmony. In the future, as our populations mix more and more, these isolating mechanisms should eventually disappear altogether, but in the meantime they need to be understood. If we imagine wrongly that they reflect deeper differences between the races, they will continue to cause trouble. They may be conspicuous, but they are nevertheless trivial and superficial and should be viewed as such.

Turning now specifically to the female head of hair, it is clear that her long tresses and naked face must have made a striking visual contrast. If, as I have argued, the excessive hair growth on top of her head evolved primarily as a visual display, it should come as no surprise that, over the centuries, it has been the subject of so much attention, both positive and negative. It has been shown off, concealed, styled, cut, trimmed, extended, straightened, waved, put up, let down, coloured and decorated in a thousand different ways. It has been everything from woman's crowning glory to the cause of strict religious taboos. No other part of the female body has been subjected to such an incredible range of cultural variations.

Before examining these variations in more detail, it is worth taking a closer look at the individual hairs themselves. There are about 100,000 on each human head. Blondes have finer hair and compensate for this by having a slightly larger number than average – usually about 140,000. Brunettes have about 108,000 hairs, while redheads, who have the coarsest hair, possess only 90,000.

Typically, each hair grows for about six years. It then goes into a three-month resting phase before falling out. At any one time, 90 per cent of the hairs are actively growing, while 10 per cent are resting. In a full human lifespan each hair papilla therefore grows about 12 hairs, one after the other. Unlike many other mammals, humans do not have seasonal moults. Our scalp hair is of the same thickness in all seasons.

On average, each individual hair grows 13 cm (5 inches) a year.

But among healthy young adults this is increased to 18 cm (7 inches) a year. So, for them, each scalp hair would, if left untrimmed, grow to be just over a metre (3½ feet) long before dropping out. This far exceeds anything found in other primates and is one of the truly unique features of the human species.

In rare cases there is a curious exception to this rule. Instead of falling out after six years, the hairs simply keep on growing longer and longer, until they reach ground level. In some instances they go on far beyond this and some women have grown hair so long that they can stand on it. One young American woman possessed hair that was over 4 metres (13 feet) in length, but even her extraordinary achievement was beaten by a Chinese woman whose world record hair was measured at almost 5 metres (16 feet). It is as though such a momentum has been developed in the genetic trend to evolve longer human hair that occasionally it runs away with itself, creating super-haired individuals.

Even without these extremes, it is clear that, with so much head hair to play with, the ever inventive human being was soon going to be tempted to start experimenting with different shapes and styles. We know, from some of the oldest Venus figurines, that this has been going on for at least 20,000 years. Stone Age carvings have been found which clearly show several distinct hairstyles, including elaborate partings down the middle of the head and, in one case, with the addition of braided hair thrown over the right shoulder.

Looking back at earlier historical periods, it is possible to see how the predominant styles have slowly changed, with each epoch showing hair fashions that are highly characteristic of their time. In the modern era, with the arrival of professional styling salons and global communication systems, the speed of these fashion changes has dramatically accelerated.

Today, in the twenty-first century, there are so many competing influences that a single theme no longer exists. With individuality the order of the day, there are more hairstyles on display than ever before. The urge to imitate celebrities still creates short-lived mini-trends, but there are so many role-models to choose from that no clear leader emerges to cry out 'This is the dominant hairstyle of the early twenty-first century'. The short, efficient hair of the female

politician, the long flowing hair of the pop star, the carefully 'dishev-
elled' hair of the Hollywood actress, the wild spiky hair of the rebel,
all these and more can be seen side by side in our morning newspapers. And even to give crude labels to these competing styles in
this way is a stereotyping that is unjustified because, within these
types, there are countless minor variations.

This is not the place to list all these inventive variations in detail,
but it is important to record that, over the centuries, there have
been a small number of major 'female hair strategies'. These are
related, not to the whims of fashion, but to the basic possibilities
of what can be done with female head hair. Some of these strategies have vanished into history and today seem very strange. Others
are still with us.

The simplest strategy of all is to adopt the Natural Look. In this,
the woman wears her hair down, loose and natural at all times,
whether in public or private, and both on special social occasions
and for everyday use. She may wash, brush and comb her hair, but
does not attempt to dress or style it in any way. Although this is
the most basic of all strategies, it is now comparatively rare. It may
still be found in unsophisticated societies or in cultures where
simplicity has become a social doctrine. Poverty may spawn it, but
even where there is no money to spend on hair products or professional hair care, native women are fond of dressing their hair.
Twisting, plaiting and braiding costs little or nothing and helps to
pass the time.

For women whose lives involve strenuous physical work – in the
fields or in the factory, for example – a Practical Look is introduced.
The hair is tied back for convenience, to prevent it falling over the
eyes or becoming entangled. When the woman is not labouring at
her daily tasks, she unties it and lets it fall loose. This has been a
popular peasant strategy in past times and is still used by many
women today who, although no longer engaged in hard physical
labour, find that screwing the hair back in a ponytail can be a useful
device to control untidy hair, both in the workplace and the home.

For the majority of women, especially those living in urban societies, the natural and the practical solutions have never been enough.
For centuries they have adopted the Styled Look in which the hair

is dressed in some way – clipped, shaped, coloured, waved, straightened, layered, streaked, or decorated. This is the usual strategy, especially in those countries where hairdressing salons are popular, but it is forbidden in countries where strict religious rules are applied and displays of female beauty are taboo.

Two of the major strategies involved in styling the hair are to enlarge it or to reduce it. Enlarging the hair increases the strength of the visual display of whatever modification has been chosen. It makes the woman seem taller and more conspicuous. A favourite way of accomplishing this has been to wear a wig of some kind.

The wearing of false hair is a strategy that is at least 5,000 years old. In ancient Egypt it was the custom for high-ranking females to shave the head completely and then to wear an ornate wig in public. Roman ladies did not shave their heads, but they too enjoyed wearing fancy wigs as a status display. Their form of showing off led to an unpleasant trend when it became fashionable to insist that the hair from which their wigs were made should be taken from the heads of conquered people whose countries had been defeated by the Roman army – an ancient Roman version of scalping one's enemies.

Fancy wigs were banned by the Church in medieval times, but reappeared in the Elizabethan era. This was largely because the primitive cosmetics of the period did so much damage to hair and skin that a massive cover-up was required. But the fancy wig was not to reach its zenith until the eighteenth century, when exaggeration upon exaggeration occurred, until the hair displays of fashionable women came to exceed anything seen either before or since. Some of their wigs were over 75 cm (30 inches) in height and elaborately decorated. Doorways had to be raised to allow them to pass through. The seating in carriages had to be specially lowered. Special bed supports had to be designed so that women could lie down and rest while still wearing their enormous wigs. At the Paris Opera, wigs were permitted only in the boxes – their presence elsewhere would have obscured the stage. Never has any other hair strategy made such an impact on society. It was an example of a special kind of conspicuous consumption. Because of the huge cost of making and maintaining the wigs, the husbands of the wig-wearers had to be unusually generous in funding the fashion. As a result, their wives' hair displays have been referred to as an example of 'surrogate consumption' – a way of showing off how rich their husbands were.

The only woman who could bring an end to this outrageously extreme fashion was Madame Guillotine, who chopped off the aristocratic wig-wearer's heads. After the French Revolution the use of fancy wigs never made a full recovery. There were moments when it resurfaced briefly in one form or another – as with the Fun Wigs of the 1960s, made out of synthetic materials and in a range of bright, artificial colours – but the great days were over. In more modern times, where wigs have been worn they have usually been so realistic as to deny their own existence.

Some women (especially those whose real hair is thinning with age) will never appear in public without a realistic wig in place. A number of famous celebrities also adopt this strategy, not because of hair problems, but as a matter of convenience. Even if their own hair is in good condition, it is often easier for them to don a wig than to waste valuable time having their hair styled. The great advantage of this is that a whole series of elegant wigs can be kept groomed and in perfect condition in the absence of the wearer.

Returning to the strategy of hair enlargement, a remarkable example from the recent past is the Big-Hair Look that became popular in the 1980s. In this, instead of wearing a wig, the owner's real hair was made to look as voluminous as possible. The appearance of increased bulk was obtained by 'blow-drying upside down, teasing out, mousse-ing and then copiously spraying'. The gravity defying result was wryly described by one critic as 'one of the architectural marvels of our time'. Sometimes referred to as Dolly Parton Hair (after the American country singer), this bouffant style was especially popular in small-town America and the Southern States, where the motto 'the higher the hair, the nearer to God' was often heard. One of the reasons for its popularity was that its great size made large facial features seem smaller and therefore more attractive. It was also extrovert and cheerfully assertive, making the wearer appear more confident. To its opponents, however, it was brash and vulgar, and nothing more than a compensation for inadequacies. And it had one major shortcoming – it may have been a blatant female advertisement, but also happened to be anti-sexual – because men could not run their hands through it, ruffle it, or gently stroke it.

More recently, a more sophisticated form of hair enlargement has been favoured. Hair extensions have been added to the natural hair to make it appear much longer. This is done either to enjoy a temporary change from a short hairstyle, or when the natural hair fails to grow as long as the woman wishes. Modern hairdressing skills have made it almost impossible to detect the presence of these hair extensions, although some of them are visibly, deliberately false and act almost like a kind of partial wig.

The second major styling strategy is to reduce the natural hair in some way, either by removing some of it, or by tightly restricting it. In its least extreme version it takes the form of adopting a severe, controlled hairstyle as a social display on special occasions, but with it worn down, loose and natural for everyday use and in private. In recent decades many women wish to appear casually 'free and easy' for most of the time, but will go to extra trouble for very special occasions, such as funerals, weddings, and grand events or celebrations. In order to give themselves a sense of social standing and discipline they usually put their hair up or restrain it in some way. This display says to the onlooker, 'I am important, I am serious, do not be familiar with me.'

Some women go a step further, never letting their hair down in public. They keep it closely confined in a tight bun or some other constricted style at all times until in the privacy of the home. This is what might be called the 'governess' or 'head-mistress' strategy. Women who need to impose their authority on others can increase their air of control and power by clamping their hair as tightly as possible onto their skulls. This de-feminizes them and rids them of any air of casual relaxation or personal freedom. Their hair is so neat that it cannot be ruffled, so tightly arranged that it cannot be stroked. This makes them appear both literally and metaphorically unruffled and renders them unapproachable and untouchable.

Some women have chosen to have their hair bobbed so short that it can no longer be 'tied back', 'put up', or 'let down'. What little is left of it may still hang loose, but it no longer requires holding back to facilitate physical labour, nor can its style be changed to suit different social contexts. The flappers of the 1920s were the first to adopt this style as a fashion statement and it reappeared in the 1960s through the work of stylist Vidal Sassoon.

Clearly, the intended message of the Short Look is that the women wearing it are active and carefree. They reduce their hair to an elegant but highly abbreviated, tomboy statement rather than a fussy feminine display. The drawback, however, is that, although in principle this made sense, in practice the short styles of the 1920s and 1960s proved rather difficult to keep in good condition after leaving the salon.

The Short Look resurfaced again in the 1970s when, in a more severe form, it became a common feminist strategy, usually as an assertive display in which women in the workplace were seeking to be treated with more respect by their male colleagues. By the 1990s the short styles had softened. They now had a more feminine touch. The hair strategy of the post-feminist businesswoman was saying 'I am still disciplined, but I no longer have to give up my femininity to be a top player in this world.' As a softer short hairstyle, the 1990s look walked the tightrope between being too belligerently butch on the one hand, or too decoratively ornate on the other. Its aim was to combine polished control with a sense of sexy freedom. This became the new challenge for the professional hairstylist in the West as the twenty-first century dawned.

In a more drastic form of hair reduction, some women have taken the step of close-cropping their hair. This removes all 'natural looseness', all the time, even when they are in private. For beautiful women this is a defiant statement which says 'look at me, I do not need pretty hair to make me attractive'. As such it can be seen as a display of vanity. It is also the statement of a rebel, one who ignores convention and refuses to follow the hair trends of fashionable or conformist women. Those women who dislike it see it as a deliberate attempt at self-advertisement by the use of shock tactics. Men may simply feel threatened by it, having been robbed of the quest for the softly flowing locks they dream of caressing.

Certain women have taken the even more extreme step of close-shaving their heads to remove all trace of head hair completely. In some cultures, shaving the female head had been carried out as a punishment. In others it has been a mark of slavery, or voluntary subordination to a deity. In still others it has been required of all women, at special mourning ceremonies. Among the ancient Phoenicians, failure to adopt the shaven head of mourning meant that the women concerned had to offer themselves as temple prostitutes. In recent times a French fashion designer persuaded all his models to shave their heads to demonstrate that modern women should not be 'imprisoned by their own hair'. For men, this shaved look has nearly always been unappetizing and (from Joan of Arc to the female punk rockers) has had little or no sex appeal, being a complete denial of all that is sensual about long female hair.

Because of its capacity for arousing men, the exposure of an expanse of female hair – in any style – has sometimes been prohibited. Covering or obscuring the hair is required to eliminate its potentially erotic signalling. The mildest form of this puritanical 'cover-up' is the wearing of some sort of headgear. The demand that women wear a hat or a scarf when entering a Catholic church is a reminder of the times when they were required to obscure their hair completely when attending Christian services. A modern remnant of this ancient practice is the social convention of wearing hats on formal occasions, such as weddings and funerals.

In strict religious communities, past and present, women are required to cover their heads completely at all times in public, only uncovering their hair in the privacy of their homes, when no strangers are present. In societies where Islamic law is rigidly imposed, for example, this look is ever-present. Even accidentally displaying a small wisp of hair, from under the traditional head covering, when walking in the street, has led to women being beaten by male religious officials. The stricter communities of the Christian Church have also imposed rules concerning female hair covering. In the past, these rules often applied to devout wives, whose hair could not be seen in public, and even today they still apply to nuns.

One extraordinary example of religious hair concealment is found today in New York, among Orthodox Jewish communities. Women in these communities must cover their hair totally in public and may only let their natural hair been seen by their husbands in the privacy of the bedroom. Despite this, women in these communities wish to live the lives of typical New Yorkers and solve their dilemma in an ingenious way. They have highly realistic wigs manufactured at great cost to look exactly like their own real hair. When they place such a wig, called a *sheitel,* over the top of their own hair, their appearance hardly changes. A casual onlooker would find it hard to tell whether they were wearing the wig or not. In this way the religious rule is obeyed without sacrificing personal image.

It is clear that hair invites experimentation more than any other part of the female body. This is because it is easy to change, these changes can be made quickly and they are not permanent. When the hair grows out, new styles can be tried. Above all, hair is highly conspicuous and even the smallest alteration in hairstyle is immediately obvious to even the casual onlooker.

In the symbolism of female hair there is a simple dichotomy – a contrast between the long, free-flowing, strokeable, natural hair and the short, severe, tightly dressed hair. Long, loose hair has been seen as symbolizing lack of restraint, sexuality, freedom of spirit, peaceful rebellion and creativity. Short, tight hair has been associated with discipline, self-control, efficiency, conformity and assertiveness. These are obviously crude over-generalizations, but it is surprising how well they fit the facts in many cases. The great joy of hair for the human female, however, remains that it is always available for her to express her personal style and her individuality, as well as her general mood. Providing the grim world of sexist religious practices does not interfere, she can use her hair as a wonderfully expressive appendage to present herself to the world.

In addition to the wide range of shaping and styling options that exist, there is also the question of modification of hair colour. The natural variation, from very dark to very pale, is, like skin colour, an adaptation to the climatic conditions of the environment. Each colour, whether black, brown, red or yellow, has its own adaptive significance and its own special appeal. It is therefore surprising to find that, when women decide to change the colour of their hair, there is one choice that completely dominates all the others. Of every hundred women who take the step of radically altering their hair colour, it would be fair to say that over 90 per cent of them choose to become blonde. At first sight this is mystifying. Why on earth should so many dark-haired women want to look like fair-haired Scandinavians, when so few Scandinavians ever want to dye their hair brown or black? It clearly has nothing to do with climate. Nor does it have anything to do with race, since the majority of Caucasians are dark-haired. So what is the special appeal of blonde hair, an appeal so strong that it has led to the bizarre situation that there are now more artificial blondes in the world than real ones?

Part of blonde appeal lies in the fineness of the hair. The unusual thinness of the blonde strands makes them genuinely softer to the touch and therefore more sensuous in moments of intimate body contact. Under stroking fingers, or against the male cheek, the soft-ness of the hair echoes the softness of rounded female flesh. So in that sense, blondes *are* more feminine than redheads or brunettes.

Indeed, femininity of blondness extends over the whole body. The blonde has a fine, soft fuzz where the brunette must wield razor or depilatory. In particular, the blonde's armpits and pubis are more delicately hirsute. Her pubic hair's soft silkiness contrasts strikingly with the more aggressive bushiness of the brunette's. In moments of extreme intimacy she therefore has a slight advantage over darker-haired women.

If it is argued that this softness of blonde hair is what makes so many darker-haired women lighten their tresses, it should be pointed out that any advantage gained is by association only. The bleaching of dark hair does not make it any finer, or softer. It merely *looks* finer.

There is, however, another advantage in being blonde, and this one depends purely on visual signalling: being blonde creates a more juvenile image than being dark-haired. And such an image projected by an adult human helps to increase their sex appeal, transmitting intense 'take care of me' signals. The reason blondness suggests youth is that for a large section of humanity, babies are blonder than their parents, and the 'baby blues' and 'blond locks' become indelibly associated with childhood.

Needless to say, this is good news for hair dyers and wig makers. From the empires of the ancient world to the salons of baroque Europe, generation after generation of dark- or mousy-haired women streamed to their establishments for the latest styles and potions, intent on making themselves a little, or a lot, blonder than nature had intended. Virtually since the dawn of recorded history, the blonding of women has been a major industry.

Some of the measures taken to meet the social demands of blondness were hazardous and even, on occasion, lethal. The ancient Greeks employed a pomade of yellow flower petals, a potassium solution and coloured powders that 'veiled the hair' to gain the sexy blonde look. Roman ladies dyed their hair with a German soap, specially imported from the north, but were more likely to take the easy way out by wearing a blonde wig. These early wigs were made of the fair human hair of the northern Europeans the Romans had conquered in their great expansion. The fashion became so widespread that the Roman poet Martial mocked it with the lines:

> The golden hair that Galla wears
> Is hers – who would have thought it?
> She swears 'tis hers, and true she swears
> For I know where she bought it.

More and more tricks were used to bleach the hair as the centuries passed. Ashes of plants, nutshells, elderberries and vinegar sediment were popular in the early days. Saffron was scrubbed into the hair with great vigour. Boiled egg yolks and wild honey were tried, aided by prolonged exposure to strong sunlight. Elizabethan women powdered their hair with gold dust or, more economically, applied the scrapings of rhubarb steeped in white wine. Sometimes they took the risk of dipping their hair in oil of vitriol or alum water. For some, these chemical treatments solved the problem of unwanted dark hair so effectively that they went completely bald and were forced to wear blonde wigs for the rest of their fashionable lives.

The recipes became more and more complicated and demanding. By 1825 a learned treatise called *The Art of Beauty* informed its readers of a formula they should brew if they wished to display flaxen hair: Boil up a quart of lye; add half an ounce of briny celandine roots and turmeric; two drachms of saffron and lily roots, and a drachm each of flowers of mullein, yellow stechas, broom, and St John's wort. The resulting concoction was to be applied regularly to the scalp.

It is clear that for year after year and century after century, many a socially aware female was prepared to go to great lengths to acquire the desirable tones of blondness. But like many fashionable conceits, the bleaching of hair inevitably developed a secondary association with over-exaggerated display. Even in Roman times its appeal did not always remain that of the unblemished virgin. The artificiality of wigs and dyes decreased the symbolic value of pale colouring. At one stage it became synonymous not with innocent femininity, but with professional sexuality: it became the sign of the whore.

Roman prostitutes were carefully organized. They were licensed, taxed, and actually required by law to wear blonde hair. The third wife of the Emperor Claudius, the wild nymphomaniac Messalina, was so excited by the idea of sudden, brutal sex with strangers that she would sneak out at night clad in a whore's wig and prowl the city. So violent was her lovemaking that it is rumoured she frequently dislodged her blonde hairpiece, returning to the royal precincts in all too recognizable condition.

Other Roman ladies of fashion were soon imitating her, and the lawmakers were impotent to stem the trend. Their blonde-wig-whoring law was ruined, but the element of wickedness and abandon by now associated with blondness was to survive down the centuries, repeatedly re-surfacing as an opposing strand in contrast to the image of fair-haired virginal innocence. Usually the distinction was made that true blondes were angels and fake blondes were promiscuous. The fact that the artificial blondes went to much trouble to look appealing meant that they had sex very much on their minds; the extreme and therefore imitation blonde became the archetypal good-time girl, the blonde bombshell, the swinger, the dolly bird, the bimbo. Each generation had its name for her, and each generation had its super-blondes.

In the wake of the First World War, the platinum blonde appeared on the scene. When Jean Harlow died at the age of twenty-six in 1937, she had already set in train a long succession of blonde movie stars – golden girls who have continued to dominate the screen from that day to this. The vast majority of the great female personalities to come out of Hollywood have been blonde – usually by design rather than genetics. Some have gone to great lengths to perfect their blondness, Marilyn Monroe even painfully bleaching her pubic hair to match her platinum tresses. Most were faithful to the ancient association between the sun and the gold in their hair – they were cheerful and warm, life-giving, and life-enhancing. They often came a cropper, but that too was part of their natural appeal – their child-like blonde vulnerability.

In defence of brunettes, a commentator in the late 1960s remarked: 'If a man is serious about a girl he wants her to be natural. Anything artificial does not appeal to a serious thinking man . . . Generally speaking he prefers a blonde for a mistress and a brunette for a wife. Brunettes have more integrity.'

3. The Brow

The brow region of the face plays a major role in body language. An eighteenth-century expert on facial expressions said of the forehead that 'of all parts of the face it is the most important and the most characteristic'. To some people today this statement might seem surprising, because so much attention is paid to eye make-up and lipstick, with the result that the eyes and the mouth tend to dominate the female face and overshadow other parts of it. But despite this, it is doubtful if anyone has ever enjoyed a face-to-face conversation without unconsciously employing forehead signals in the form of eyebrow movements and skin wrinkling – actions that are vitally important in indicating changing moods.

Before examining these signals and the ways in which the female brow differs from the male, it is worth asking why we have a brow region at all. If you take a close look at a chimpanzee's face, side by side with a human face, the forehead difference is striking. In the case of the ape, the brow region is almost non-existent. In humans, the brow rises vertically above the eyes as a great expanse of naked skin, conspicuously decorated with two small eyebrow patches. In stark contrast, the chimpanzee's hairline comes right down to the eyebrows, which themselves are almost hairless. In fact, the brow region of the ape is the complete opposite of that in human beings.

When you look at a chimpanzee's face, or that of any other ape, the impression is that they have huge, prominent eyebrow ridges – protruding bony crests that help to protect their eyes from damage – while we have lost ours. But this is an illusion. If you touch the bony region just above your eye sockets, with your fingertips, you can feel the thickening of the skull that is still there to protect you. That it is less obvious in humans is not caused by its disappearance but by the fact that the forehead above it has expanded dramatically to house our much bigger brains. The human brow ridges have not vanished, they have simply been engulfed. The brain of the chimpanzee has a volume of only about 400 cubic centimetres, whereas the human brain has a volume of 1,350 cubic centimetres – more than three times the size. It was this expansion of the human brain, especially in the frontal region, that gave us a 'face above our eyes'.

Possessing a new, exclusively human area of skin above the eyes provided our ancient ancestors with an additional zone for visual signalling. This is made possible by the fact that the forehead skin, although stretched tight over the front of the skull, is not completely immobile. It is capable of slight movement – small but clearly visible. This movement is easy to detect because, when the skin shifts, it creates visible wrinkle patterns. Also, more importantly, the human face has retained two small patches of hair on its otherwise smooth-skinned forehead. These hairs, known technically as superciliary patches, but more usually referred to as eyebrows, act as 'markers' that help to make the skin movements even more conspicuous at a distance.

Once, it was thought that the main duty of the eyebrows was to prevent sweat and rain from dripping into the eyes. Although they may be of minor assistance in this way, acting as deflecting 'gutters' at the bottom of the forehead, their main role today is undoubtedly to act as signalling devices, transmitting information to companions about second-by-second alterations in the state of mind of their owners.

Studying the female face in all its many moods, it is clear that there are six distinct eyebrow actions, each linked to a particular emotional state. They are as follows:

Eyebrows Lower. This action, the frown, is not strictly vertical. As the eyebrows move downwards they also move slightly inwards, coming closer together. This has the effect of squeezing the skin between them and throwing it into short, vertical folds. The number of these folds varies from individual to individual and each adult has a characteristic 'frown pattern' of one, two, three or four lines. Frequently they are asymmetrical, with the lines on one side of the inter-eyebrow space (an area known as the glabellum) being longer or stronger than the other.

The horizontal wrinkle lines on the forehead tend to be smoothed out by the action of lowering the eyebrows but they may not disappear completely. Part of the aging process in the human animal involves an increasing fixation of temporary expression lines. Skin creases which, in the young, appear and vanish again with each changing mood, become permanently etched into the skin surface as the years go by. The strength of a frown line on a non-frowning face is a fair indication of the sum total of past frowning performed by that individual.

Eyebrow lowering occurs in two quite different kinds of situation that can be crudely labelled as aggressive and protective. In aggressive contexts the action covers a wide range of intensities, from mere disapproval or self-assertive determination right through to annoyance and violent anger. In protective contexts the action occurs whenever there is a threat to the eyes.

At moments of danger the lowering of the brows over the eyes is not protective enough, however, and in such instances there is also a raising of the cheeks from below. Together these two actions provide the maximum level of eye protection possible, while still keeping the eyes open and active. This 'screwing up' of the eyes is typical of a wincing face that anticipates physical attack, or of an over-illuminated face exposed to such strong light that the eyes are beginning to suffer from it.

This protective 'screwing up' also occurs frequently during laughter, crying and at moments of intense disgust, indicating that these conditions can also perhaps be looked upon as kinds of over-exposure.

It is the eye protection function that explains the ancient origin of the lowering of the brow region. The use of this lowering in aggressive contexts appears to be secondary, based on a need to defend the eyes from the retaliatory attacks that the aggressive mood can be expected to provoke. We often think of the frowning face as a 'fierce' face and therefore not one that should be connected with self-protection, but this would appear to be an error. It may be fierce, but it is not so fearlessly fierce as to be unconcerned about the need for self-protection of those crucially important organs, the eyes. The truly fearless face of aggression, by contrast, displays a pair of staring, unfrowning eyes, but this is a comparatively rare occurrence, since overtly hostile acts are seldom safe from some kind of retaliatory response.

Eyebrows Raise. Like the last movement, this action is not strictly perpendicular. As the eyebrows are raised they move slightly outwards, away from each other. This has the effect of stretching the skin between them and flattening out the short vertical frown-folds there. At the same time, however, the whole of the forehead skin is squeezed upwards, creating a pattern of long horizontal crease-lines. These are roughly parallel with one another and number four or five in most cases. Sometimes there are as few as three or as many as ten, but it is difficult to be precise because the upper and lower lines are usually fragmentary. Only the middle lines run right across the brow in most instances.

This is the 'furrowed brow' of popular speech and is usually thought of as belonging to a 'worried' individual. Its true range however, is much wider than that. It has been described by various authors as signifying: wonder, amazement, surprise, happiness, scepticism, negation, ignorance, arrogance, anticipation, querying, incomprehension, anxiety and fear. One music critic famously commented that a particular opera star 'was a singer who had to take any note above A with her eyebrows'. With all these different interpretations, the only way to understand the significance of this movement is to look backwards towards its ancient origin.

Brow raising is a pattern we share with other primate species and for them, like us, it appears to have started out as a vision-improving device. The pulling up of the forehead skin and the raising of the eyebrows has the immediate effect of increasing the overall range of vision. To use a well-known phrase, it is an 'eye-opener'. It increases visual input.

Among monkeys it seems to be an emergency reaction, brought into use whenever the animals are confronted by something that makes them want to flee. It only occurs, however, if, at the same time, there is something else in the situation that stops them from fleeing. This 'something else' can be a number of things. It can be a conflicting urge to attack, or a burning curiosity to stay and look at the thing that is so frightening, or any other tendency to stay put which conflicts with the urge to flee and blocks it.

If we now apply this concept of 'thwarted escape' to the human context it fits remarkably well. Humans and monkeys behave in much the same way. The worried person with the furrowed brow is essentially someone who would like to escape but for some reason cannot do so. The laughing individual with the same brow-wrinkling expression is also slightly alarmed. There are telltale elements of body withdrawal in this posture. The laughter may be genuine but whatever is being laughed at is also rather disturbing. This is not uncommon. Much humour takes us to the brink of fear and makes us laugh only because it does not push us over. The arrogant person with stiffly arched eyebrows would also like to escape from the surrounding crassness. In popular speech, we give such an individual the appropriate name of 'highbrow'.

When this expression is compared with the lowering of the eyebrows, a problem emerges. Suppose we see something frightening in front of us: we can either lower our brows to protect our eyes, or raise our brows to increase our range of vision. Both will be helpful, but we have to choose. The brain has to assess which is the more important demand and instruct the face accordingly. If we look at monkeys, we see that during their most aggressive threats their brows are lowered; during their rather scared threats their brows are raised; and during moments of beaten submission their brows are lowered again. The position is much the same in humans.

When human beings are very aggressive and might provoke an immediate retaliation, or where they are beaten and fearful of imminent attack, they sacrifice improved vision and protect their eyes with lowered eyebrows. When they are slightly aggressive but also very scared, or where they are in any state of conflict that does not appear to be in imminent danger of turning into a physical attack, they sacrifice eye protection for the tactical advantages of being able to see more clearly what is happening around them and raise their eyebrows.

Once these two actions have developed their primary roles, they can be used as indicators in quite mild contexts. We put them to work as contrived signals. People may raise their eyebrows deliberately, even when they are not worried, simply in order to signal to someone 'How worrying for *you*'. But such refinements and modifications would not be possible were it not for the original primitive significance of the actions.

As with frown lines, the arched creases caused by raising the eyebrows can become indelibly fixed as we grow older. The skin of our foreheads becomes scarred with traces of all the grimacing we have done over the years. If we are frequently nervous or anxious, our repeatedly raised forehead skin becomes permanently marked with thin arched lines. The elasticity of the skin declines as we age and, like a piece of creased paper that we try to flatten out, the forehead refuses to regain its completely smooth, juvenile condition, even at times of relaxation and calm.

The female forehead that becomes lined in this way is a telltale sign that its owner is no longer young. It also suggests an over-anxious personality. 'Old and nervous' is not what an image-conscious female wishes to transmit, and so she has to take some kind of action to repair the damage, or at least to conceal it.

Heavy make-up can help, but it can only go so far. A thick fringe of hair can act as a useful cover-up, until a gust of wind dislodges it. For those women who make a living from their looks, more drastic action is needed. For many years now, the surgical option has been the face-lift. It is drastic but effective, pulling the skin so tightly over the skull that it is incapable of showing even the tiniest wrinkle.

A more modern alternative for eliminating wrinkles, available since the 1990s, is a course of *Botox* brow injections. These have the effect of paralysing the forehead so that it is incapable of any movement, no matter how emotional its owner may be. *Botox* is, in fact, a poison, a protein neurotoxin produced by the *Clostridium botulinum* bacterium. It is injected straight into the muscles that cause the wrinkles, inactivating them for a period of three to five months. In this type of cosmetic treatment it is used in such small quantities that there is little or no danger, although it has yet to be approved as safe by official medical bodies. Despite this, it is said to have become the second most popular form of cosmetic surgery in use at the present time.

The problem with these severe solutions is that they leave the forehead almost *too* smooth, incapable of showing any emotion at all. This can create a mask-like appearance – a young face, but a rigid face. A more perfect medical solution remains to be found.

Eyebrow Cock. This is a mixture of the previous two actions, with one eyebrow being lowered while the other is raised. It is not a particularly common expression – many people find it hard to perform.

The message given by this action is as intermediate as the expression itself. Half the face looks aggressive while the other half looks scared. For some reason, this contradictory response is observed far less frequently in adult females than in males. The mood of the eyebrow cocker is usually one of scepticism. The single raised eyebrow acts rather like a question mark in relation to the other, glaring eye.

Eyebrows Knit. The eyebrows are simultaneously raised and drawn towards each other. Like the last action, this is a complex one made up of two elements taken from the Lower and the Raise. The inward movement is taken from the Eyebrows Lower action, producing short vertical creases in the narrowed space between the eyebrows. The upward movement is taken from the Eyebrows Raise action, producing horizontal creases across the forehead. The knitting of the brows therefore produces a double set of skin wrinkles.

This is the expression associated with intense anxiety and grief. It is also observed in some cases of chronic, as opposed to acute, pain. A sudden sharp pain gives rise to the Face Wince reaction, with its lowered brows, but a dull prolonged pain is more likely to produce the Eyebrows Knit posture. A good example of this is the standard expression employed in advertisements for headache medication.

In origin, this action appears to be an attempt by the brow to respond to a double signal from the brain. One message says 'raise the brows' and the other says 'lower them'. Different sets of muscles start to pull in opposite directions. The first set manages to pull the eyebrows up a little, but the second set, although trying to tug them down and in, only manages to pull them in towards each other.

In some cases, but by no means all, the inner ends of the eyebrow are pulled up further than the outer ends, resulting in the 'oblique eyebrows of grief'. This exaggerated form of the knitting action is most marked in those who have seen more than their fair share of tragedy. If women with less tragic histories try to force their eyebrows up into the oblique position, they may have little success, even though they may be able to feel their brows trying to move into this special posture. Theoretically it should be possible to tell how much misfortune there has been in a woman's past life simply by checking the ease with which she manages to adopt the 'oblique brows' posture.

Eyebrows Flash. The eyebrows are raised and lowered again in a fraction of a second. This brief upward flick of the eyebrows is an important and apparently universal human greeting signal. It has been recorded not only in Europeans from many areas, but also in people from as far afield as Bali, New Guinea and the Amazon basin, sometimes in instances where there have been no European influences. In each case it has the same meaning, indicating a friendly recognition of the other person's presence.

The Eyebrows Flash is usually performed at a distance, at the beginning of an encounter, and is not part of the close proximity displays, such as handshaking, kissing and embracing, that follow. It is often accompanied by a head toss and a smile, but may also occur by itself.

In origin it is clearly a momentary adoption of the Eyebrows Raise posture of surprise. Combined with the smile it becomes a signal of pleasant surprise. The extreme brevity of the action, no more than a split second, indicates that the surprised mood is quickly banished, leaving the friendly smile to dominate the scene.

As already mentioned, the raising of the eyebrows has an element of fear in it, and it may seem strange that such a factor should play a part in a greeting between friends. But every greeting, no matter how amicable, involves an increase in social unpredictability. We simply do not know how the others are going to behave or how they may have changed since we last met them. This inevitably tinges the encounter with a small, fleeting element of fear.

In addition to its role as a greeting signal, the Eyebrows Flash is frequently employed during ordinary conversational speech as a 'marking point' for emphasis. Each time a word is stressed strongly, the eyebrows flick up and back again. Most of us do this occasionally but with some individuals it becomes particularly frequent and exaggerated. It is as if we are saying 'these are the surprise-points' in the verbal communication.

Eyebrows Shrug. The eyebrows are raised, momentarily held in the upper position and then lowered again. It is the brief 'hold' of the eyebrows in the raised position that distinguishes this action from the Eyebrows Flash of greeting and emphasis.

This is the eyebrow element of the complex shrug reaction, which also involves a special posture of the mouth, head, shoulders, arms and hands. Almost all the elements of this compound display can also occur separately, in isolation, or in conjunction with one or two other elements. The Eyebrows Shrug, although it may sometimes occur entirely on its own, is usually accompanied by a Mouth Shrug – rapid and momentary turning down of the mouth corners. This combination – what might be called the Face Shrug – frequently occurs in the absence of the other shrug elements.

Unlike the Eyebrows Flash, therefore, this action is typically linked with a 'sad' mouth rather than a 'happy' one. This should give it the meaning of a mildly unpleasant surprise, which is frequently the way it is used. If, for instance, two people are sitting together and a third one nearby does something socially 'uncomfortable', one of the two intimates may give the Eyebrows Shrug to the other to indicate surprised disapproval.

The Eyebrows Shrug is also often seen as an accompaniment to speech in certain individuals. Nearly all of us, when talking animatedly, make repeated, small body movements to stress what we at saying. At each point of verbal emphasis we add a visual emphasis. With most people it is the hands or the head that keep moving with each stressed point, but some individuals favour their eyebrows. As they speak, their eyebrows are repeatedly shrugged in time with the special points they are making. This is typical of the speech of the chronic 'complainer', who seems to be perpetually surprised by the vagaries of life, but it is by no means confined to this particular personality type.

Leaving the question of eyebrow movements and turning to their anatomy, there is an important gender difference: female eyebrows are thinner and less bushy than those of males. This difference has led to many forms of 'improvement'. Female eyebrows have been made super-female by artificially increasing their thinness and smallness.

This has been done for centuries, using a variety of techniques such as shaving, waxing, plucking and painting. Originally the excuse given was that these procedures helped to ward off evil; later they were supposed to protect the body from disease and, in particular, to prevent blindness; later still, they were done to enhance personal beauty. In all instances the underlying urge was to make the eyebrows look more exaggeratedly feminine.

In the twentieth century, the peak of eyebrow plucking came in the interwar period, in the 1920s and 1930s, when 'the eyebrow pencil was in every vanity bag, available in five bewitching shades'. After the thickness of the eyebrows had been reduced by the use of tweezers, the eyebrow pencil was employed to emphasize the thin arched line of hair that survived.

For some eyebrow shapers, the use of tweezers is considered too crude. At the moment of plucking, the metal ends of the tweezers could easily break the hair and this means that it will quickly re-grow. A preferred method is 'threading', in which a tiny thread is tied around each hair before it is pulled out. This ensures that the root of the hair is also removed. This method is especially popular in Asia and the Middle East.

If a particular woman felt that her eyebrows were in an unbecoming position on her forehead, she could, of course, remove them altogether and paint in totally artificial lines in a new position to suit her taste. When this step was taken, the new eyebrows nearly always appeared above the true position, giving her face less of a frown. There is a late eighteenth-century saying that 'Eyebrows gently arched harmonize with the modesty of a young virgin' and it is certainly true that artificially raised eyebrows do give a more innocent, childlike, 'wide-eyed' look. A woman with naturally low-set eyebrows can look so sinister that she has sometimes been referred to as having 'witch's brows'.

The precise shape of the modified eyebrows has varied from era to era and from individual to individual. Great care has often been taken to sculpt the eyebrows in a way that both obeys the fashion of the period and yet, at the same time, is a perfect match for the rest of the face. One eyebrow virtuoso insists that the ideal shape 'is formed from a two-thirds upward sweeping curve and a one-third downward sweeping curve', although this may have to be adapted to fit certain individual faces. Such are the subtleties of eyebrow aesthetics.

Perhaps the strangest example of false eyebrows comes from England in the early eighteenth century. At that time, fashionable eyebrows were also shaved and then replaced, but it was the nature of the replacement that was so bizarre. The eyebrow 'falsies' of the day were made of mouse-skin. Swift records this strange fashion with the words: 'Her eyebrows from a mouse's hide, Stuck on with art on either side.'

Because these attentions were all concerned with improving feminine appearance, it followed that leaving the eyebrows in their natural, unplucked condition was frequently a statement of a strong 'non-sexual' nature. Women working in conditions where they were expected to suppress their sexuality were also expected to leave their eyebrows alone. In the 1930s there was a hotly debated case involving a London hospital where a matron had refused permission for a nurse to pluck her eyebrows. A complaint was made to the effect that this was interfering with personal liberty, but the matron's decision was upheld by the London County Council. Hospital patients were, as a result, spared the erotic stimulation of the sight of a delicately plucked eyebrow, as they lay in their sickbeds. (Someone who would have been pleased with this ruling is the

Prophet Mohammed, one of whose sayings contains the phrase 'May God curse the women who . . . pluck their eyebrows.')

Finally, a mention must be made of the female face that displays a dramatic 'monobrow' or 'unibrow'. In this condition the two eyebrows are joined across their inner ends to create one long, continuous brow-patch. It is not common, and where it does exist it is seldom left untrimmed. Any woman who is born with this feature will usually go to great pains to pluck out the offending hairs that cover the glabellum at the top of her nose. There are several reasons for this. First, the extra hairiness of the brow region is unduly masculine. Second, there is an 'animal' quality to having hair where usually there is none. Third, if the central hair is left there it gives the impression of a permanent frown. And fourth, there is a centuries-old superstition that any woman possessing a unibrow must be a vampire.

Together, these unfortunate associations are enough to have any fashionable woman searching for her tweezers. To retain the sinister unibrow, a woman would have to be 'above fashion'. One such woman did exist in the twentieth century – the famous, bisexual Mexican artist Frida Kahlo. For her, the hairy glabellum became a personal trademark and she reproduced it faithfully in all her self-portraits. Described as 'hovering above her piercing dark eyes like a bird in flight', her jet black unibrow was also likened to a major landmark. As one critic put it: 'Frida Kahlo may have been an interesting and creative woman, but she only had one eyebrow. It stretched from one side of her face to the other, like the Great Wall of China and, like that wall, it was probably visible from the moon.'

It is amusing that these extreme reactions are all caused by the presence of a few dark hairs at the top of Kahlo's nose, and it is remarkable that such a small extension of the human eyebrow patches should produce such a powerful response. We take the eyebrows so much for granted that it is only when something odd happens to them that we sit up and take notice.

Apart from Frida Kahlo's idiosyncratic unibrow, there has only been one time in recent years when heavy female eyebrows were considered acceptable and, for a while, even became popular. That was in the 1980s, when the feminist movement had reached the stage where women decided that looking more like men was a good way to compete with them. This was the era that saw the young Hollywood star, Brooke Shields, appear in movies with what were described as 'caterpillar brows'. They may not have joined in the middle, like Kahlo's, but they were as heavy as any man's, and gave her a fiercely determined, highly assertive look. Since then, as women have concluded that it is much more appropriate and satisfying to succeed *as women,* rather than as pseudo-males, their eyebrows have returned once more to the trimmed and arched condition that has been in favour for centuries. As Shakespeare put it in *The Winter's Tale*: 'Yet black brows, they say, become some women best, so that there be not too much hair there, but in the semicircle or a half-moon made with a pen.'

4. The Ears

Female ears have never been treated well. They have either been ignored or they have been mutilated. The powder and paint that has been so lovingly applied to the female face has passed them by. While the meticulously decorated face has taken centre stage, the ears have been ignored, frequently hidden beneath female hair. And when they have emerged from hiding, they have been considered fit only for drilling to create jewellery supports. On the rare occasions when ears have been the subject of cosmetic surgery, the treatment has been only to make them even more inconspicuous – as when slightly protruding ears are pinned back more tightly against the sides of the female head. But before examining in more detail the cultural abuses heaped on the long-suffering female ear, it is worth pausing to consider the biology and anatomy of this part of the female body.

The visible part of the human ear is a rather modest affair. During the course of evolution it has lost its long pointed tip and its mobility. Its fine, sensitive edges have gone too, curled over into a 'rolled rim'. But it should certainly not be dismissed as a useless remnant.

The main function of the external ear remains that of a sound gatherer – a flesh-and-blood ear trumpet. We may not be able to prick up our ears like other animals, or to twist and turn them when seeking the direction of a sudden noise, but we are still capable of detecting the source of a sound to within 3 degrees. What humans have lost in ear mobility we have made up for with head mobility. When a deer or an antelope hears an alarming sound, it raises its head and twists its ears this way and that. When we hear such a sound, we turn our heads, and it works almost as well.

Although our ears feel rigid at the sides of the head, they do retain a ghost of the movements they once boasted. If you strain the muscles of the ear region tight, while looking in a mirror, you can just see a trace of the protective movement, as your ear tries to flatten itself against the side of your head. Animals with big mobile ears nearly always flatten them when fighting – trying to keep them out of harm's way – and we still do this automatically when we tighten the head skin in moments of panic, even though our ears are already flattened in their normal resting posture.

The shape of our external ears is important in delivering undistorted sound to our eardrums. A person unfortunate enough to have their ears sliced off would find themselves far less efficient as listeners. The ear canals and their eardrums would form a 'resonant system' with some sounds being emphasized at the expense of others. The seemingly haphazard shape of the ear – its curving folds and ridges – is in reality a special design preventing any distortion of this kind.

A minor function of our ears is temperature control. Elephants flap their huge ears when they are overheated and this helps to cool the animals down. There is a profusion of blood vessels near the surface of the skin and heat loss by this route can be important to many species. For us it may only play a trivial role in thermoregulation, but it has become a social signal. If a woman overheats in a moment of emotional conflict, her ears may go bright red. This ear blushing has been the subject of comment since ancient times. Nearly 2,000 years ago, Pliny wrote: 'When our ears glow and tingle, someone is talking of us in our absence.' And Shakespeare makes Beatrice ask the question 'What fire is in mine ears?' when she is being discussed by others.

Finally, our ears appear to have acquired a new erotic function with the development of soft fleshy lobes. These are absent in our nearest relatives and appear to be a uniquely human feature, evolved as part of our increased sexuality. Early anatomists dismissed them as functionless: 'a new feature which apparently serves no useful purpose, unless it is pierced for the carrying of ornaments'; but recent observations of sexual behaviour have revealed that during

36

intense arousal the earlobes become swollen and engorged with blood. This makes them unusually sensitive to touch. Caressing, sucking and kissing of the lobes during lovemaking acts as a strong sexual stimulus for many women. In rare instances, according to Kinsey and his colleagues at the Institute for Sex Research in Indiana, a woman may even reach orgasm as a result of stimulation of her ears.

At the centre of the external ear is the shadowy 'ear hole' that leads to a narrow canal about an inch long. The canal twists slightly, giving it a design that helps to keep the air inside it warm. This warmth is important for the proper functioning of the eardrum at its inner end. The eardrum itself is an extremely delicate organ and the canal not only keeps it snugly warm but also protects it from physical damage. The price we must pay for this protection, however, is the presence on our bodies of a deep recess that we cannot clean with our fingers. We can groom the rest of our bodies with comparative ease, freeing ourselves from dirt and small parasites, and we can snort, sniff or blow our noses to cleanse those other open recesses, the nostrils; but if an invader enters one of our ear canals, we are in trouble. Attempts to remove the irritation with a thin stick can easily damage the eardrum and we clearly need some special defence against intrusions of this kind. Evolution has provided the answer in the shape of hairs to keep out larger insects, and earwax to defeat smaller creatures. The orange-coloured earwax has a bitter taste which is repellent to insects. This wax is produced by 4,000 tiny ceruminous glands that are highly modified apocrine glands – the kind which produce the strong-smelling sweat in the regions of the armpits and the crotch.

This is not the place to tell the detailed story of the interior parts of the ear. Briefly, the sound vibrations that strike the eardrum are converted into nervous impulses for transmission to the brain. The eardrum is incredibly sensitive, capable of detecting a vibration so faint that it displaces the surface of the drum only a thousand-millionth of a centimetre. This displacement is then transmitted through three strangely shaped bones (the hammer, the anvil and the stirrup) in the middle ear, which amplify the pressure 22 times. The enlarged signal is then passed on to the inner ear where an odd, snail-shaped organ filled with fluid is activated. Vibrations are set up in this fluid that impinge on hair-like nerve cells. There are thousands of these nerve cells – each one tuned to a particular vibration – and they send their messages to the brain via the auditory nerve.

The inner ear also contains vital organs of balance, three semi-circular canals, one of which deals with up-and-down movements, one with forward movements and one with side-to-side movements. The importance of these organs grew dramatically when our ancestors first stood up on their hind legs and adopted bipedal locomotion. An animal standing on four legs is reasonably stable, but vertical living creates an almost non-stop demand for subtle balancing adjustments. Although we take these balancing organs for granted, they are in fact more crucial to our survival than the parts of the ear that deal with sounds. A deaf person can survive more easily than one who has completely lost the sense of balance.

One of the unfortunate aspects of our sense of hearing is that it starts to go into decline as soon as we are born. The human infant can detect sound wave frequencies from 16 cycles a second up to 30,000. At adolescence, the upper limit has already dropped to 20,000 cycles a second. By the age of sixty this has declined to about 12,000 and the upper pitch that we can detect continues to fall further and further as we become more elderly. For the very old, it becomes a problem to listen to conversation in a crowded room, although they may still be happy enough listening to a single voice in a quiet place. This is because, with their greatly narrowed range of hearing, they find it hard to distinguish between different voices when several are speaking at once.

Modern hi-fi systems are efficient up to frequencies of 20,000 cycles a second, and it is a galling thought for a middle-aged woman who has just spent a large sum of money to install such a system that the only members of her family who will be able to appreciate its full range will be her young children. She herself will be lucky if she can detect its output at anything above 15,000 cycles a second.

Our ears have one serious weakness in relation to volume of sound. We evolved, like other species, in a comparatively quiet world, when the loudest sounds we heard were roars and screams.

There was nothing louder to damage our sensitive eardrums and so we developed no special protection against very loud sounds. Today, thanks to our infinite ingenuity, we have thundering machines, high explosives and a whole variety of super-sounds which can very easily damage our hearing. Clearly, our ears serve as a reminder to us all that we now live in a very different world from the one in which we evolved.

Returning to the external ear, it has long been argued that it is possible to identify every individual by their ear shape. In the last century it was suggested that this feature could be used to detect criminals, but a rival method – fingerprinting – won the day and ear-typing was forgotten. It remains true, however, that it is impossible to find two people with precisely the same ear details. Thirteen ear zones have been labelled and two of them deserve special mention.

The first is the fleshy lobe. Apart from variations in its size, this has one major classificatory feature. Each of us has either 'free' lobes or 'attached' lobes. Free lobes hang down slightly from their point of contact with the head; attached lobes do not. A doctor who went to the trouble of examining 4,171 European ears discovered that 64 per cent of them had free lobes and 36 per cent attached lobes.

The second is a small bump on the rim of the ear, called Darwin's Point. It is present in most ears, but is often so slight that it is hard to detect. If you feel down the inside edge of the folded rim of your ear starting at the top, you will come across it about a third of the way down the height of your ear. It feels like a slight swelling, not much more than a swollen pimple, but Darwin was convinced that it was significant remnant of our primeval days, when we had long pointed ears that could be moved about freely, searching for small sounds. In his own words, these 'points are vestiges of the tips of formerly erect and pointed ears'. Careful studies have revealed that they are present in a conspicuous form in about 26 per cent of Europeans.

It is variations in details like these that make the ears a suitable subject for criminal identification, but the use of fingerprinting has now reached such an advanced state that it is doubtful whether ear shapes will ever be needed. Unhappily the only people to make elaborate studies of ear zones at present are latter-day physiognomists with their romantic claims of character and personality identification from the reading of facial proportions. Their fanciful comments, which lost all credibility early in the twentieth century, astonishingly resurfaced in the 1980s, when it was possible to read that a large ear is the ear of an achiever; that a small, well-shaped ear belongs to a conformist; and that a pointed ear is that of an opportunist. These and hundreds of other such 'readings', often going into great detail, are an insult to human intelligence and their popularity in the late twentieth century is hard to understand.

Criminologists studying facial details have reported that ear design can never be predicted from facial design. If you see a rounded face and an angular face it is impossible to predict which will have the more rounded or angular ear shapes. Somatotyping experts do not quite agree with this. They claim that endomorphs (the more roly-poly among us) and ectomorphs (the more bony and angular) do have different types of ears. Endomorph ears are characterized as lying flat against the head and having the lobe and the pinna (the flap of the ear) equally well developed. Ectomorph ears, by contrast, have the pinnae projecting laterally and better developed than the lobes. Perhaps the explanation for this disagreement is that the criminologists are only considering the head region, while the somato-typers are examining the whole body shape.

Symbolically, the ear has been given several roles. Because it is a flap of skin surrounding an orifice it has inevitably been viewed as a symbol of the female genitals. In Yugoslavia, for instance, a slang expression for the vulva is 'the ear between the legs'. In some cultures mutilation of the ears was employed as a substitute for female circumcision. In parts of the Orient young girls at puberty were forced to go through an initiation ritual in which holes were bored in their ears. In ancient Egypt the punishment for an adulteress was the removal of her ears with a sharp knife – another example of the ears as genital substitutes.

Because the ears were seen as female genitals in many different cultures, it is not particularly surprising to discover that certain exceptional individuals were born through the ear. Karna, the son of the Hindu sun god, Surya, is said to have emerged from this organ, and it has been pointed out that, technically speaking, Karna's ear-delivery meant that his mother, Kunti, enjoyed a virgin birth. In some legends it has also been claimed that Buddha was born from his mother's ear.

In the satirical works of François Rabelais, published in 1653, Gargantua also entered the world in this unusual fashion. When Gargamelle was about to give birth to him, 'the child sprung up and leapt, and so entering the hollow vein, did climb by the diaphragm even above her shoulders, where that vein divides itself into two, and from thence taking his way towards the left side, issued forth at her left ear'. The author admits that this is hard to believe, but defends himself by pointing out that there is nothing in the Bible to contradict such a form of birth and that, if God wished it, 'all women henceforth should bring forth their children at the ear'.

A completely different form of ear symbolism sees this organ as representing wisdom. This is because it is the ear that hears the word of God. This has been given as the excuse for pulling the ears of children when they are naughty, the idea being that the activation of the ear will wake up the intelligence lying dormant there.

Some of these strange superstitions were to lead to the ancient custom of piercing the ears for earrings. This primitive form of mutilation has proved remarkably tenacious and is one of the few types of artificial deformity that has retained a widespread popularity in the modern world. Today most women who have their ears pierced do so for purely decorative purposes, with no understanding of what this once meant. In ancient times there were several explanations:

Because the Devil and other evil spirits are always attempting to enter the human body to take it over, it is necessary to protect all the orifices through which they might gain access. The wearing of lucky charms on the ears was thought to have been the best way to guard against an earful of demons.

Because ears are the seat of wisdom, it follows that the very wise have very big ears, especially earlobes. Heavy earrings that pull down the lobes and make them even longer must therefore increase natural wisdom and intelligence. A study of Hindu, Buddhist and

Chinese sculptured figures from earlier days reveals that, if they were important royal persons, they always possessed elongated earlobes.

Other early beliefs were that wearing earrings would cure bad eyesight, or would act as a protection from drowning.

Over a long period of time these original, varied reasons for donning earrings have been forgotten. In the modern era, nearly all earrings, both tribal and urban, are purely decorative and have been worn solely for reasons of beauty or status. In tribal cultures, where long earlobes have been fashionable, the process of mutilation usually starts out in infancy, with tiny children having holes bored in their earlobes. These small holes are then gradually enlarged, year by year, so that the ears stretch further and further downwards. By puberty, only the girls with the longest ears are considered to be beautiful. The really gorgeous ones have to have ears down to their breasts. If, in the process, the long hanging loop of ear-flesh snaps under the strain of repeated tugging and the weight of heavy ear ornaments, a girl's beauty is instantly ruined. In some cultures she is then considered to be too ugly to marry.

Surprisingly, examples of these extreme elongations of the female ears are found scattered all over the world. The custom appears to have arisen independently, time and again, from places as far apart as Borneo and Brazil, Africa and Cambodia. Among the Trobriand Islanders, if a girl had dared to ignore this custom, or she would have been ridiculed as 'having the ears of a bush-pig'.

In some tribes there is a special festival associated with the ritual of piercing the earlobes of the young girls. In certain cultures, the heavy ornaments hanging from the stretched earlobes of a married woman may not be taken off until her husband dies. They are then removed as a sign of mourning at the funeral ceremony.

The size of the earlobe ornaments is sometimes astonishing. In one tribe, as many as 50 brass rings, 10 cm (4 inches) in diameter, are hung from each ear. In another, heavy copper rings are added until the total weight pulling down each extended lobe is as much as 1 kg (2.5 lb). In yet another, full-size jam pots, or food cans, begged from Westerners, are inserted into the gaping earlobe holes.

In earlier centuries, the Western world was shocked and horrified by these excessive forms of mutilation. John Bulwer, writing in 1654, devoted a whole chapter of his book *A View of the People of the Whole World* to attacking 'Auricular fashions, or certain strange inventions of people, in new-moulding their ears', castigating women who 'think it a great comeliness to have their ears most shamefully bored', making a hole in it and 'putting lead into it, which with its weight so extends it, that it hangs down to the shoulders, the hole so big that you may put your arm through it.' To Bulwer and his epoch, any attempt to improve or modify the human form was an insult to God.

Such disapproval did little to halt these tribal customs. They were too much a part of cultural history to be abandoned lightly. In a few cases foreign influences may have eroded the more extreme forms of mutilation but in many of the more remote societies they still survive undiluted even in the twenty-first century.

Despite its often extravagant fashions, the Western world has never displayed anything to compete with the extended earlobes of those tribal societies. The most extreme examples we can offer are to be found in the brief flowering of punk rock in the 1970s. Seeking to outrage, the punks searched for bizarre objects to thrust into their clumsily drilled lobes. Large safety pins were clear favourites, but chains carrying everything from razor blades to electric light bulbs were also used by the shock troops of the new wave. Their mood was too impatient, however, for them to bother with the prolonged and gradual stretching of the earlobes found in tribal groups.

Later, with the dramatic increase in body piercing that started to flourish towards the end of the twentieth century, Western female ears could be seen with multiple earrings. Instead of one hole, the ear was drilled again and again, all around its perimeter, so that a whole series of earrings could be attached to it.

For the vast majority of women, however, the ears today are decorated in a simple way, with easily removable, stud or drop earrings attached through a single, small hole, or simply clipped on. Unlike tribal earrings they are not worn continuously, but are often changed daily to match whatever else is being worn at the time. Some women own only a few pairs, but others become addicted to acquiring large numbers, the record-holder (according to the *Guinness Book of Records*) being an American woman from Pennsylvania who has amassed a collection of 17,122 pairs. If she wore a different one each day, it would take her nearly half a century to get through them all.

5. The Eyes

For many centuries the female eyes have been the focus of a great deal of attention. Eye make-up is known to have been employed for over 6,000 years. In ancient Egypt, black cosmetics were used to colour the eyelids, and the Roman satirist Martial, writing in the first century, made the waspish comment: 'still you wink at men under an eyelid which you took out of a drawer that same morning'. Eyelids, eyelashes and the skin around the eyes have been the subject of endless subtle variations in colour, shade and embellishment in every major civilization in the history of the world. Eye shadow, eyeliners, eyelash curlers, false eyelashes and coloured contact lenses have all been employed to enhance the female eye region. But before taking a closer look at these improvements on nature, what about the natural eye itself?

The eyes are the dominant sense organs of the human body. It has been estimated that 80 per cent of our information about the outside world enters through these remarkable structures. Despite all the talking and listening we do we remain essentially visual animals. In this we do not differ greatly from our close relatives, the monkeys and apes. The whole primate order is a vision-dominated group, with the two eyes brought to the front of the head, providing a binocular view of the world.

The human eye is only about 2.5 cm (1 inch) in diameter and yet it makes the most sophisticated television camera look like something from the Stone Age. The light-sensitive retina at the back of the eyeball contains 137 million cells that send messages to the brain, telling it what we are seeing. Of these, 130 million are rod-shaped and concerned with black-and-white vision; the remaining 7 million are cone-shaped and facilitate colour vision. At any one moment these light-responsive cells can deal with one-and-a-half million simultaneous messages. Because it is so complex, it is hardly surprising that the eye is the part of the body that shows the least growth between birth and adulthood. Even the brain grows more than the eye.

At the eye's centre is the black pupil – the aperture through which light passes to fall on the retina. The pupil increases in size with weak light and decreases in strong light, controlling the amount of illumination falling on the retina. In this respect the eye does act much like a camera with an adjustable diaphragm, but it also has a curious override system. If the eye sees something it likes very much the pupil expands rather more than normal, and if it sees something distasteful it shrinks to a pinprick. It is easy to understand the second of these two responses because further contraction of the pupil's aperture would simply reduce the illumination of the retina and 'damp down' the distasteful image. The increased pupil dilation that occurs when we see something attractive is harder to explain. This must interfere with the accuracy of our vision by letting too much light flood on to the retina. The result must be a hazy glow rather than a sharp balanced image. This may be an advantage for young lovers, however, when they gaze deeply into each other's dilated pupils. They may benefit by seeing a slightly fuzzy image bathed in a halo of light – the very opposite of a 'warts-and-all' image.

In earlier centuries, the courtesans of Italy would put drops of belladonna in their eyes just before receiving a visitor. This had the effect of dilating their pupils massively, which made them more appealing because it gave the false impression that they loved what they saw (even if it was only the bloated, ravaged face of an elderly roué).

Around the pupil is the muscular, coloured iris, the contracting disc responsible for the changes in pupil size. This task is performed by involuntary muscles, so we can never deliberately or consciously control our pupil size. It is this fact that makes pupil expansion and contraction such a reliable guide to our emotional responses to visual images. Our pupils cannot lie.

The colour of the iris varies considerably from person to person, but this is not due to a variety of pigments. Blue-eyed people do not possess blue eye pigment; they simply have *less* pigment than others, and this gives the impression of blueness. If you have a dark brown ring around your pupils, this means that you have a generous amount of melanin pigment in the front layers of your irises. If the melanin here is less and the pigment is largely confined to the deeper layers of the iris, then your eyes will be paler, ranging from hazel or green to grey or blue as the pigment decreases. Violet colouring is due to blood showing through.

Brightly coloured eyes in human beings are therefore something of an optical illusion. They indicate a loss of melanin and seem to be part of the general 'paling' of the body that occurs as one moves away from the equator towards the less sunny polar zones. This effect is most striking when one compares the babies of white people with those of darker-skinned races. Almost all white babies are blue-eyed at birth. Dark-skinned babies are dark-eyed. Then, as they grow older, most white offspring gradually develop the melanin pigment on the front of the iris, making their eyes darker and darker. Only a very small percentage fail to do this and retain their 'baby blues'.

Covering the pupil and the iris is a transparent window, the cornea, and around it is the region we refer to as 'the whites of the eyes', technically called the sclera. It is this non-optical part of the human eye that is its most unusual feature. Uniquely, parts of the whites of our eyes are visible to onlookers. Most animals have circular 'button' eyes. The same is true of lower primates, but with most monkeys the skin around the eyes is slightly pulled back, left and right, to give the eyes 'corners'. These eyes are still nearer to a circle in shape than an oval; but the trend goes a step further in the apes, where the eyes are more elliptical, approaching the human shape. Even here no 'whites' are visible, the exposed area on either side of the iris matching its dark brown colour. In humans the whiteness of these same areas makes them highly conspicuous. The effect of this small evolutionary change is that during social encounters shifts in gaze direction are easily detected, even at a distance.

Surrounding the visible part of the eye, the eyelids, fringed with curved eyelashes, have greasy, shiny edges. The greasiness is cause by the secretions of rows of tiny glands, visible as minute pinpricks just behind the roots of the eyelashes. The regular blinking of these lids moistens and cleans the cornea. The process is aided by the secretion of tears from the tear gland, tucked in under the upper eyelid. The liquid is drained off through two small tear ducts – also visible as pinpricks, but bigger ones, on the edges of the eyelids. They are positioned at the nose end of the lids, one in the upper and one in the lower. The two ducts connect up into a single tube that carries the 'used' tears down into the interior of the nose and away. When an irritation in the eye or intense emotions make the tear glands produce tears more quickly than the ducts can drain them off, we weep. The excess tears spill over on to our cheeks and we wipe them away. This is the second unique feature of the human eyes, for we are the only land animal that frequently weeps with emotion.

Between the two tear ducts, in the corner of the eye next to the nose, there is a small pink lump. This is the remnant of our third eyelid and it now appears to be completely functionless. In many species it is an organ of some value. Some use it as a 'windscreen wiper', blinking sideways to clean the eye; others have coloured ones that they flash as a signal; still others have completely transparent ones they can use as natural sunglasses. Diving ducks go even further, having specially thickened transparent ones that they pull over their sensitive corneas when swimming under water. If only our primeval ancestor had been more aquatic, our sub-aqua pleasures today might have been greatly enhanced.

The eyelashes, which provide us with a protective fringe above and below the eyes, have one exceptional feature: they do not become white with age like other head and body hairs. Each eye has about 200 of them, more on the upper lid than the lower, and each lash lasts between three and five months before falling out and being replaced. Eyelashes have the same lifespan as the hairs of the eyebrows.

Another form of eye protection occurs in Orientals, who possess a flap of skin called the epicanthic fold that lies over the upper eyelid and gives their eyes their characteristic 'slant'. This fold is present in the human foetus in all races, but is retained into adulthood only by the Eastern branch of the human family. A few Western babies are born with the eye fold still present, but it gradually disappears as the nose narrows and changes shape with advancing age. Among the Oriental peoples the epicanthic fold seems to have been retained as part of a general adaptation to cold. The whole face is more fat-laden, flatter and better able to cope with icy conditions, and the extra fold of skin above the eyes helps to shield this delicate area in an extreme environment.

The shape of the Oriental eye is undeniably appealing, but many women in the Far East do not see it that way and today their hospitals are full of young, eye-bandaged figures subjecting themselves to the surgeon's knife and having their epicanthic folds removed so that their eyes have a more Western appearance.

There are small gender differences in the eye. The female eye is very slightly smaller than that of the male, and it shows a higher proportion of white. In many cultures the tear glands are more active in emotional females than in emotional males, but whether this is due to cultural training that requires the males to be less emotionally demonstrative or is a biological difference of a more basic kind is hard to say. It does appear to be a remarkably widespread difference for it to be simply the result of social training.

A word about the tears themselves: they are not only lubricants for the exposed surface of the eye, they are also bactericidal. They contain an enzyme called *lysozyme* which kills bacteria and protects the eye from infection.

Poor eyesight must have been a curse for many of our remote ancestors, not only because of the lack of precision in obtaining visual information but also because the permanent strain of trying to see with defective vision causes severe headaches and migraines. The curse remained for people of the earliest civilizations and with the invention of writing it became acute, many elderly scholars having to employ young people to read for them.

Seneca, the Roman connoisseur of the art of rhetoric who lived at the time of Christ, seems to have been the first person to attempt to solve this terrible problem. It is said that despite poor sight he managed to read his way through the libraries of Rome by using a 'globe of water' as a magnifying glass. This ingenious solution should have led to an early development of eyeglasses, but it failed to do so. It was not until the thirteenth century that the English philosopher Roger Bacon recorded his observation that 'If anyone examines letters or other minute objects through the medium of crystal or glass . . . if it be shaped like the lesser segment of a sphere, with the convex side towards the eye, he will see the letters far better and they will seem large to him.' He went on to say that such a glass would be useful to those with weak eyes, but again there was no rush to develop this boon to human vision. Towards the end of the century, in Italy, true spectacles for reading did at last appear, though it is not clear whether they were influenced by Bacon. In 1306 a monk in Florence gave a sermon that included the following phrase: 'It is not yet 20 years since the art of making spectacles, one of the most useful arts on earth, was discovered . . .' At about the same time Marco Polo recorded seeing elderly Chinese using lenses for reading, so it is clear that by the fourteenth century the move towards a wide use of eyeglasses had begun in earnest. In the fifteenth century special lenses for correcting short-sightedness appeared, and in the eighteenth century Benjamin Franklin invented bifocals. The first successful contact lenses were made in Switzerland in 1887.

This brief history of eyeglasses is of more than medical interest because it also changed the appearance of our eyes. The shape of the spectacles became part of the facial expression of the wearer. A heavy upper rim became a super-frown, making the owner look more fierce and domineering. A wide circular rim produced a wide-eyed stare, as though the curve of the rim represented arched eyebrows. There was no deception, as in the case of subtle make-up. The spectacles were clearly not part of the face and yet it was impossible not to be influenced by their lines, just as an eye mask alters the whole expression of the wearer.

The effect of dark glasses is especially dramatic. Telltale eye movements, made conspicuous by the whites of the eyes, as mentioned earlier, provide a constant source of information during social encounters, but dark glasses effectively eliminate that information.

Darting eyes, shifty eyes, inattentive eyes, over-attentive eyes, dilated eyes, all are hidden from the companions of someone wearing 'shades'. They can only guess at what is taking place behind the mask of the sunglasses.

What are they missing? Imagine a social gathering of a group of women. Precisely what do their eye movements tell us? In any such gathering subordinates tend to look at dominant figures and dominants tend to ignore subordinates, except in special circumstances. For example, if a friendly submissive individual enters a room her eyes will dart this way and that, checking on all those present. If she spots high-status, dominant individuals she will keep a persistent watchful eye on them. Whenever a joking remark or a controversial statement is made or a personal opinion is expressed the subordinate's eyes will flick in the direction of the dominant person to assess any reaction. The dominant figure typically remains aloof during such exchanges and hardly bothers to look at the subordinates during general conversation. But if she fires a straight question at one of them she does so with a direct stare. The individual on whom she fixes finds herself unable to return this stare for any length of time and during most of her reply looks elsewhere.

This is the situation where a clear pecking order operates and where certain individuals have control over others and wish to exercise it. When friends of equal status meet, the eye movements are rather different. Here everyone uses 'subordinate' eye movements even though they are not subordinate. This is done because the simplest way of demonstrating friendliness with body language is to display non-hostility and non-dominance. So we are watchful of our friends, treating them with our eyes as if they are dominants. When they speak or are active we look at them; when we speak and they watch us we look away and glance at them only briefly from time to time to check their reactions to what we are saying. In this way, each of two friends will treat the other as the powerful one and thus make each other feel good.

If a dominant female wishes to ingratiate herself with someone she can do so by deliberately adopting the friendly body language of an equal. When addressing an employee or servant of some kind, she can manipulatively switch on an attentive eye, hanging on the underling's every word. Such devices are rarely used by dominant individuals outside special contexts (such as election campaigns).

Prolonged eye-to-eye staring occurs only at moments of intense love or hate. For most people in most settings a direct stare that is held for more than a few moments is too threatening and they quickly look away. For lovers there is such total mutual trust that they can hold each other's gaze without even a twinge of fear. As they stare into each other's eyes they are unconsciously checking the degree of pupil dilation. If they see deep black pools they know intuitively that their feelings are reciprocated. If they see tiny pinprick pupils they may start to feel uneasy, sensing that all is not well in their relationship.

Turning from lovers to haters, the staring eyes of an angry person are strongly intimidating. In earlier days, when superstitions were rife, it was believed that supernatural beings were watching over human events and influencing their outcome. The fact that these divine powers or deities were *watching* meant that they must have eyes. Since they had to watch over so much it was supposed that they must have many eyes and be all-seeing. Where good gods were concerned this was clearly a great advantage to human beings because benign deities could be protective. But there were also bad gods, demons and devils – evil spirits with evil eyes – and a look from them could spell disaster.

A belief in the power of evil eyes became widespread and survives even today in some parts of the world. The evil eyes became the Evil Eye, a malicious, harmful and even deadly influence that could strike down a victim without warning. If its glance fell on you something terrible would happen. Sometimes an ordinary woman became possessed, against her will, of the Evil Eye, and everyone she looked at suffered in some way soon afterwards. Many lucky charms and talismans were employed to protect people against these threats, usually worn on a costume, kept in a handbag, or hung up in the house. Some of these protective devices operated on the principle that an intensely sexual image would distract the Evil Eye and keep it preoccupied.

Astonishingly, with this concept in mind, many Christian churches in medieval Europe displayed stone-carved images of female genitals above their doors, to stop 'the evil ones' from entering the buildings. To intensify the image, the genitals are usually shown being held open by a pair of hands. Not surprisingly, most of these carvings were removed or hidden during the pious Victorian era, but a few still survive to this day. A much better survivor is the lucky horseshoe, also placed on a building to bring good luck. Were it generally known that, in this protective context, the horseshoe originally also symbolized the female genitals, it too might start to disappear.

Since the Evil Eye's worst deeds were thought to be caused by envy it was important not to lavish praise on anyone who might be vulnerable. A mother might be horrified if a stranger praised her new baby, for example, and would have to hang a lucky charm on the infant's cot to defend it, or carry out some other protective ritual. Even today, especially in the Mediterranean region, these superstitious precautions are still taken very seriously by many people.

Turning from the imagined eyes of evil spirits to the real eyes of the human female, there are many visual messages to be read in her changing expressions:

Eyes Lower. Lowering the eyes formally is sometimes used as a modesty signal. It is based on the natural behaviour of subordinates who dare not look at their superiors, but it is not random in its direction. The primly modest 'flower' does not cast her eyes to left or right but only down at the ground. There is the suggestion of a bow in this action, or of the lowering of the head in submission.

Eyes Raise. Raising the eyes is also sometimes used as a deliberate signal. If they are held in the up position for a while, the expression is one of 'a pretence of innocence'. Performed today only in jest, this eye movement is based on the idea of looking up to heaven as a witness of the claimed innocence.

Eyes Glare. Glaring eyes are often employed by a mother when trying to subdue children while remaining silent. The glare is a complex version of the stare. The eyes fixate the 'victim' with frowning eyebrows but widely opened eyes. This is a contradiction because the opening wide of the eyes normally goes with raised eyebrows, so these two parts of the face have to work against each other. For this reason it is not an expression that is held for any length of time. During the glare the upper eyelids press upwards so hard that they almost disappear behind the descending brows, and the demarcation line of the glaring eye is provided by the brow skin, not the eyelid. This gives a strange apparent shape to the eyes that is unmistakable. The message of the glare is one of surprised anger.

Sidelong Glance. This is used to steal a look at someone without being seen to do so. It is also used as a deliberate signal of shyness when it becomes a sign of coyness. 'I am too frightened to look straight at you, but I can't help staring' is the message here, and the popular phrase 'making sheep's eyes at someone' has been coined to describe this action.

Eyes De-focus. De-focusing the eyes occurs when we are very tired or when we are daydreaming. Someone who wishes to signal that they have something special to daydream about (a new lover, for instance) may deliberately stare out of a window or across a room with de-focused eyes as a way of impressing companions.

Eyes Wide. Widening the eyes to the extent that white is showing above and/or below the iris is a basic response of moderate surprise. This action increases the field of vision of the eyes and paves the way for increased responsiveness to visual stimuli. As with many of these automatic reactions of the eyes, a deliberately 'acted' version is now sometimes used to signal mock-surprise.

Eyes Narrow. Narrowing the eyes also has its deliberate version. Basically it is a protective response against too much light or possible damage, but it has a contemptuous form in which the person narrowing the eyes is plainly *not* suffering overexposure or a threat of damage. This artificially 'pained' expression implies that those present are the cause of a more-or-less permanent anguish. It is an expression of distaste – a haughty look of disdain upon the world around. The special fold of skin above the Oriental eye sometimes creates a false impression of 'haughtiness' because it makes the eye look as though it is being deliberately narrowed.

Eyes Glisten. Glistening eyes transmit an entirely different signal and one that is hard to fake (except for professional actors). The twinkling, sparkling or glistening surface of the eyes is slightly over-loaded with secretion from the tear glands caused by aroused emotions, but the feelings are not sufficiently strong to produce actual weeping. These are the gleaming eyes of the passionate lover, the adoring fan, the proud mother and the triumphant athlete. They are also the shining eyes of anguish, distress and bereavement, in fact any strong emotional condition that stops just short of crying.

Eyes Weep. Weeping itself is also a powerful social signal. The fact that we weep and other primates do not has aroused considerable interest, and it has been suggested that this difference is due to our ancestors having passed through an aquatic phase several million years ago. Seals weep when emotionally distressed and sea otters have also been seen to weep when they have lost their young. It has been suggested that copious shedding of tears is a by-product of the improved eye-cleaning function of tears in mammals that have returned to the sea.

This aquatic explanation is certainly logical enough. If man went through an aquatic phase several million years ago, stepped up his tear production in response to prolonged exposure to sea water and then returned to dry land as a savannah hunter, he might well retain his tearful eyes, exploiting emotional weeping as a new social signal. It would explain why he is the only primate to display this characteristic. An alternative explanation is that it was the dusty world of the savannah that increased tear production and that copious emotional weeping was a by-product of improved eye-cleaning. If it is pointed out that other mammals living in dusty conditions are all non-weepers when distressed, it can be argued that they all have hairy cheeks in which flowing tears would be lost. Only on the naked facial skin of the human species would the glistening teardrops act as a powerful visual signal to nearby companions.

A completely different explanation of weeping eyes is based on the idea that tears, like urine, have excretion of waste products as their main function. Chemical analysis of tears produced by distress and those produced by irritation to the surface of the eyes has revealed that the two liquids spilling down the face contain different proteins. The suggestion is that emotional weeping is primarily a way of ridding the body of excess stress chemicals, which would explain why 'a good weep makes you feel better' – the improvement in mood being a biochemical one. The visual signal of the wet cheeks of the weeper, which encourages companions to embrace and comfort the distressed individual, must then be seen as a secondary exploitation of this waste-product-removal mechanism. Once again it is hard to see how this theory can be reconciled with the absence of weeping in such animals as chimpanzees, who suffer from intensely stressful moments during social disputes in the wild.

Eyes Blink. Leaving the dramatic subject of weeping and turning to the more mundane one of blinking, there are several deliberate signals that are in use today. The ordinary blink, the windscreen-wiper action of the eyelids that cleans and moistens the corneal surface at frequent intervals throughout the day, takes approximately $1/40$th of a second. In emotional states, as tear production starts to increase, the blink rate increases with it, so that a measure of the frequency of blinking can be used as an index of mood.

Modified forms of blinking include the Multi-blink, the Super-blink, the Eyelash Flutter and the Wink:

Eyes Multi-blink. The Multi-blink occurs when someone is on the verge of tears. It is a desperate attempt to bail out the eyes before they start spilling over. Because of this it can also be used as a conscious signal of sympathetic distress.

Eyes Super-blink. The Super-blink is a single massive exaggerated blink, slower in speed and greater in amplitude than the normal blink. It is used as a melodramatic signal of mock-surprise and is employed exclusively as a contrived 'theatrical' action. The message is 'I don't believe my eyes, so I am wiping them clean with a huge blink in order to make sure that what I am seeing is really there.'

Eyelash Flutter. The Eyelash Flutter in which the eyes are rapidly fluttered open and shut is similar to the Multi-blink but involves a greater degree of eye opening, being performed with a wide-eyed 'innocent' look. It is another contrived coquettish action of a theatrical kind, employed in a 'you can't be cross with little me' context.

The Wink. The wink is a deliberate one-eyed blink that signifies a state of collusion between the winker and the person winked at. The message of the wink is 'You and I are momentarily involved in a shared act which secretly excludes all others.' Performed between friends at a social gathering it implies that the winker and her companion are privately in sympathy over some issue, or that they are closer to each other than either is to the others present. Performed between strangers the gesture usually carries a strong sexual invitation regardless of the genders involved. Because it suggests a private understanding between two people the secretive wink can be used openly as a 'tease' gesture to make a third party feel like an outsider. Whether it is used covertly or overtly the gesture is considered improper by writers on etiquette, one authority declaring that in Europe the act of winking by a woman is not 'upper class' and might be classified with 'the jab in the ribs to make a point . . .' As one well-known television presenter frequently reminds us with her farewell gesture at the end of each programme, many women find it difficult to wink convincingly, and look awkward when attempting it. For some reason as yet undiscovered (unless it be the difficulty of winking while wearing eye make-up) men find it much easier than women to perform a convincing wink.

In origin the wink could be described as a 'directional eye closure'. Closing the eye suggests that the secret is aimed only at the person being looked at. The other eye is kept open for the rest of the world, who are excluded from the private exchange.

Because the female eyes transmit so many important visual signals, it is not surprising that they have been subjected to various cosmetic improvement. By 5000 BC, eye-paint in ancient Egypt had already become quite sophisticated. Galena, an early type of kohl composed of lead ore, was employed to make black lines that exaggerated the shape of the eyelids. Specially imported malachite, an oxide of copper, was used to make the famous green make-up that was applied to the eye region as a kind of paste. This was more than merely decorative, also acting as a protection against the sun's glare. A more elite form of purely decorative eye make-up was concocted from crushed ants' eggs.

It is clear that, for the fashionable Egyptian woman of ancient times, eye make-up was both costly and time consuming and new research has revealed that, by the second millennium BC, it became even more complicated than was previously believed. In addition to the well-known black and green colours, it is now known that, 4,000 years ago, the Egyptian lady of fashion had purple, yellow, blue and three types of white available to her, thanks to some fairly advanced chemistry. Two of the whites also acted as antibiotics. In addition, the black she used was available in both dull matt and shiny forms, depending on how finely ground it was.

To apply these cosmetics to her eyes, she used round-ended, carefully carved applicator sticks made of wood, bronze, haematite, obsidian or glass. Sets of these sticks, and beautifully decorated cosmetic pots, have been found in elaborate make-up cabinets and toilet boxes dating from over 3,000 years ago.

The actual design of the embellishment of the female Egyptian eye included one strange element – a horizontal black line that extended back from the outer side of each eye heading towards the ear. This highly characteristic, decorative element had a magical significance, because it was an imitation of the eye-markings of the cat – an animal sacred to ancient Egyptians.

This obsession with eye make-up in ancient Egypt lasted for several thousand years. Even towards the end of that great civilization, Queen Cleopatra was still experimenting with novel colour combinations, painting her upper eyelids deep blue and her lower ones bright green.

Matters were rather different in ancient Greece, where respectable women were expected to display the purity and grace of a natural complexion. Despite the fact that it was the Greek language that gave us the word 'cosmetic' (from *kasmetikos*, meaning 'skilled decoration') only the Greek courtesans were able to enjoy the facial improvements of the make-up box. For them, it was acceptable for the eyelids to be emphasized with a brush dipped in incense black, and for the eyes to be underscored with kohl. Although many Greek males enjoyed the company of such women, the painted courtesans were sneered at by the puritanical authors of the day, one of whom remarked that, to see such women getting out of bed in the morning 'one would find them even less attractive than monkeys'.

The ancient Romans were less austere in this respect. Ovid, who wrote the first-ever book on the subject of cosmetics, records the fashionable use of both black eye shadow, made from wood-ash, and golden eye shadow, derived from saffron. The Roman dramatist Plautus commented that 'a women without paint is like food without salt'.

After the fall of Rome, female eye make-up virtually disappeared in Europe and was not to re-surface for many centuries. When it did so it was usually the preserve of ladies of easy virtue – Europe following the Greek traditional. It did not re-surface fully until the beginning of the twentieth century, when a backlash against the primness of Victorian values began to gain momentum. The year 1910 saw the publication of a remarkable little volume called *The Daily Mirror Beauty Book,* in which it was daringly suggested that a pencil line might be used to elongate the eyes. It also described a device for curling the eyelashes, to make them 'look like stars'.

Following the First World War, the 1920s and the 1930s saw these Edwardian beginnings blossom into a vast commercial world of mass-marketed cosmetics. The newly emancipated young women of this period were determined to decorate their bodies to suit their own tastes and to reject any interference from male authority figures. They were strongly influenced by the early cinema. Actresses appearing on the technically primitive 'silver screens' of the day were forced to emphasize their facial features to make them clearly visible to their audiences.

One film actress in particular, Theda Bara, made a major impact on the development of mass-produced cosmetics, appearing with heavy-lidded eyes that sparked a new fashion. Her eye make-up was the brainchild of Helena Rubenstein, the first great pioneer of modern cosmetics. Rubenstein had borrowed the idea of colour-shaded eyes from the French theatre and, with a knowledge of ancient Egypt, she experimented with kohl to create Theda Bara's dramatic mascara for her film role as Cleopatra.

This was the beginning of a cosmetic revolution. Within a few decades, the extremes of Hollywood had become commonplace worldwide. In the early 1960s ancient Egypt was once again to exert a massive influence on how women decorated their eyes. This time it was Elizabeth Taylor who appeared as Cleopatra. In the 1963 cinema epic, her heavily made-up eyes became an inspiration for young women everywhere and the sales of eye shadow, eyeliner and false eyelashes boomed.

By the late 1960s the defiantly artificial eye-style of the Cleopatra-look gave way to a more natural appearance, but eye cosmetics had certainly not vanished. Their supposedly natural look was, in reality, highly contrived. The blatancy of the early 1960s eye make-up was replaced by cunning subtlety creating 'the innocence of childhood'. An advertisement proclaimed that 'To the Naked Eye It's a Naked Face'. The catch was that this 'naked face' was achieved by the longest and most painstaking cosmetic procedure in the history of make-up.

Since then, eye make-up has been an ever-present element in modern female cosmetics – sometimes subtle, sometimes less so – with the skin shadows above and below the eyes, the eye-line and the eyelashes all receiving varying attention, as dictated by the temporary whims of fashion. In the Western world, at least, there is no sign of any impending restriction in this area of female 'modification'. Even in countries where religious dogma demands the subjugation of women to the extent that they must cover their faces in public, there is (if only the male religious persecutors could see it) elaborate eye make-up that is given as much attention as ever – even if it can only be enjoyed in private. As one Iranian female author put it, women may be forced to look plain by the leaders of the Islamic Republic, but 'ironically the Iranian make-up industry is booming'. Clearly, the female desire to emphasize the beauty of their eyes remains as strong today as it did in the times of the ancient civilizations.

6. THE NOSE

The nose is a very small part of a woman's anatomy but it has a significance that is out of proportion to its size. It is an inexpressive part of a woman's face, capable of little more than wrinkling itself up in disgust. Yet it has always attracted an unusual amount of attention. Its precise shape has been of great importance in judging a woman's beauty, so much so that cosmetic operations to modify female nose shape have been in great demand for over half a century. Why should this be? What is so special about this part of the female anatomy? Why does Tennyson wax lyrical about 'her slender nose, tip-tilted like the petal of a flower'?

It is all too clear why, in the evolution of our species, such features as wide childbearing hips, healthy glowing skin and ample breasts should have a powerful impact as primary signals of female appeal, but what possible evolutionary advantage could there be in the exact shape of a female nose? To understand this it is first necessary to examine the basic biology of the nose.

If we compare the human nose with the noses of our animal relatives, it soon becomes apparent that our nose, with its prominent bridge, its elongated tip and down-turned nostrils, is unique. Monkeys and apes have nothing quite like it. Those that do have a long snout also have a long face to match. We have a protruding nose on a flat face; this strange condition requires a special explanation.

Some anatomists have offered the unconvincing argument that as the human face became flatter during the course of evolution, the nose simply stayed where it was, like a large rock being exposed

61

by the receding tide. It is hard to accept this view. There is something so positive about the independence of the nose from its surrounding facial elements, that the 'projectile organ', as it has been called, must give some specific biological advantage to its owners. Several have been proposed.

The first theory sees the proud human proboscis as a resonator. Its enlarged condition is interpreted as a move in support of the ever-growing importance of human vocalization. As the voice evolved and speech developed, so, it is argued, did the nose. To illustrate this it is only necessary to try talking while pinching the nose shut between thumb and forefinger. The loss of vocal quality is dramatic. This is why opera singers are so terrified of catching a cold. But perhaps the clear human voice needs only the large sinuses – the hidden nasal cavities – to resonate efficiently? If this is the case we still need some other explanation for the protruding, external nose.

A second theory sees the human nose as a shield – a bony piece of armour helping to protect the eyes. If you place the tip of your thumb on your cheekbone, a fingertip on your eyebrow and another on the bridge of your nose, you will feel your hand pushing against three defensive protrusions surrounding the eye. This bony triangle protects the soft and vulnerable eye from frontal blows.

A third and rather fanciful idea sees the nose as a shield against water. It is argued that our ancestors may have gone through an aquatic phase several million years ago and that during those watery days our bodies adapted in a number of ways. In this view, the nose is seen as a shield against the inrush of water when diving. It is pointed out that when we jump into the water we hold our noses, but that we do not need to do this when we dive headfirst. This is true, but it seems much more likely that, if human beings did go through a prolonged aquatic phase, we would have taken the more obvious step of evolving nostril valves like a seal. It would only require a small evolutionary step to develop a nose that could shut tight under water. If we did this there would be no need to develop a long nose-tip with down-turned nostrils – and valvular nostrils would be so much more useful to an aquatic ape.

But perhaps the shape of the human nose helped it to act as a shield of a different kind – a shield against dust and wind-driven dirt. Leaving the peace of the trees and moving out on to the open plains and other more hostile environments, our remote ancestors must have encountered harsh, windy conditions where a down-turned nose would have served them well. This argument sees the nose as an air-conditioning plant faced with an increasing burden as our ancestors spread out into colder and drier regions of the earth. To understand this it is necessary to take a look inside the nose.

When the outside air is inhaled through the nostrils it is hardly ever in an ideal state for passing on to the lungs. The lungs are fussy about the kind of air they get – ideally it must be at a temperature of 35°C (95°F), a humidity of 95 per cent, and it must be free of dust. In other words, it must be warm, wet and clean, to prevent the delicate linings of the lungs from drying out or becoming damaged. The nose achieves this in a remarkable way, supplying over 14 cubic metres (500 cubic feet) of conditioned air regularly every 24 hours.

If hospital patients lose the use of their nose for some reason they will find their lungs in serious trouble within only a day or so. Attempts to produce an artificial nose for such patients have run into many difficulties, underlining the amazing engineering efficiency of the human nose.

The whole of the inner surface of the complex nasal cavities is covered with a mucous membrane that secretes about 1 litre (2 pints) of water a day. This damp surface is not static: it is always on the move, because embedded in it are millions of minute hairs called cilia. These keep beating away at a rate of 250 times a minute, shifting the mucous blanket along about half an inch a minute. With the help of gravity the mucous layer moves down towards the back of the throat, where it is swallowed. While this is happening, the air passing through these cavities gets warmer and warmer and more and more damp. The dust and dirt it is carrying collects on the mucus and is swept away. The lungs are safe for another breath.

It follows from this that as our ancestors spread out from their original, steamy tropical environment, moving into the grasslands to hunt game, the demands put upon their noses were gradually increased. For example, in a *hot moist* climate 76 per cent of the moisture comes from the outside and the nose is only asked to contribute 24 per cent. In a *hot dry* climate, on the other hand, only 27 per cent of the moisture comes from the air, while 73 per cent must come from the nasal linings. This means that to retain its efficiency, the nose of the hot savannah or desert dweller must be much taller and more prominent that that of the rainforest dweller.

Today, modern human beings whose ancestors have been inhabiting their present zones for a long period of time do have noses to match. Careful mapping reveals that it is possible to classify people by their nasal index and show how they then fall into regional groups that match up with the temperature and humidity. This does not mean that they are being classified into what are usually called the 'races' of man. For example, dark-skinned people living in hot wet areas in, say, West Africa, will have much wider flatter noses than even darker-skinned people living in the much drier grasslands of East Africa. Nose shape is simply an indication of the kind of air your ancestors breathed and nothing else.

To sum up, then, the human nose is a resonator and a bony shield which grew taller and longer as our species spread out and away from its hot moist Garden of Eden, keeping its air-conditioning function up to scratch. But there is more to the nose than this, of course, for it is also our main organ of smell and 'taste'. The smelling is carried out by two small odour-detecting patches of cells about the size of a small coin, high up in the nasal passages. These patches are made up of about five million yellowish cells that give us a much better sensitivity to fragrances and odours than we usually realize. We are capable of detecting certain substances in dilutions of less than one part in several billion parts of air. And experiments have proved that the human nose is good enough to be able to follow a fresh trail of invisible human footprints across a 'carpet' of clean blotting paper.

The female nose has a remarkable sensitivity to masculine odours. Research carried out in the 1970s identified over 200 different chemical compounds that can be found in body fluids such as sweat, saliva, skin oils and genital liquids. Amazingly, it was discovered that women who enjoy regular, frequent sexual encounters, during which, inevitably, a complex bouquet of male fragrances wafts into their nasal cavities, have a much more balanced physiology. They experience more regular menstrual cycles and fewer fertility problems – such is the power of the nose.

Mothers are also able to identify their own babies purely by their body fragrance. If, in a simple experiment, a group of mothers is lined up and blindfolded and their babies are then carried down the line, one by one, each mother is able to pick out her own offspring and distinguish it from all the others. The young women are usually surprised to discover that they can do this – it being a sensitivity that they did not realize they possessed. Once again, the ability of the human nose has been underestimated. (For the record, only 50 per cent of the young fathers were successful.)

The reason we are unaware of the high efficiency of our nose is because we have increasingly ignored and interfered with its operations. We live in towns and cities where natural fragrances are smothered, we wear clothes that sour our natural healthy body odours, and we spray our world full of scent killers and scent maskers. We even think of 'smelling' as somehow primitive and brutish – an ancient ability best forgotten and left behind. Only in certain specialized areas – the wine taster, the perfumer – is there any attempt to educate the modern nose and develop its full and extraordinary potential.

Calling the nose our main organ of taste as well as smell requires explanation. The tongue is the true organ of taste, but it is very crude in its ability. It can distinguish only four qualities – sweet, sour, bitter and salty. All the other 'tastes' of our widely varied cuisine are in fact detected not on the eager surfaces of our slobbering tongues, as we munch and chew and gulp our meals, but on the small odour-sensitive patches high up in our nasal cavities. Odour-bearing particles make their way there either directly through the nose as we bring the food to our mouths, or indirectly from the mouth itself. A meal may taste good (on the tongue) but it smells delicious (in the nose).

That, then, is the biology of the nose, but how can it help us to understand the strong link that exists between female nose shape and feminine beauty? One answer can be found in the unique, bony protrusion of the human nose – the way that the bridge of the nose sticks out on the otherwise rather flat human face. If, as has been suggested, this helped to protect the eyes from violent blows, then it follows that primeval hunting males would need more protection that primeval food-gathering females. In primitive tribes the adult females were too valuable to send off on the hunt. Adult males were more expendable, but even so, if they were to face the physical dangers of the hunt, they needed all the protection they could get. One way they acquired it was by developing heavier skulls, with thicker brows, stronger cheekbones and larger nose-bridges. Together, these gave better protection to their eyes. And the larger nose-bridges meant that, on average, male noses were going to be bigger than female noses.

In addition, the increased athleticism of the males as they pursued their fleet-footed prey meant that the nose became more important as an air-conditioning unit. Again, there was an evolutionary pressure for the male's nose to be bigger than the female's.

This gender difference created the equation: smaller nose = feminine nose. It followed that any female born with an unusually modest nose would be seen as super-feminine. Any female born with an unusually big nose would feel uncomfortable.

This was not all. There was a further influence at work, favouring the small female nose. As babies we all possess tiny, button noses. As we progress through childhood these small projections grow in proportion to the rest of the face and reach their greatest level with adulthood. So it follows that a small nose is a young nose. Add to this situation a 'cult of youth' and the consequence is clear: the smaller your nose the younger you look.

So, to be youthfully feminine it is doubly important to have a small nose. For most women this is not a problem – nature has equipped them appropriately. For some, however, there is a feeling that they have been genetically unlucky and have been dealt an unduly large, masculine nose. There are two possible reasons for this. One is that they have been unfortunate simply as part of the individual variation that occurs in all populations. The other is that their recent ancestors came from a part of the world where larger noses were a valuable adaptation to the extreme climate that was prevalent there. Noses from hot dry desert regions, like the Middle East and North Africa, are taller than usual; those from hot humid regions, like certain parts of tropical Africa, are wider than usual. If people from these regions find themselves living in other parts of the world, where the climate is more temperate, some of them will feel that their noses are not feminine enough and will wish they were smaller. Until the last century there was little they could do about it, but then the introduction of more advanced techniques in cosmetic surgery came to their rescue – the 'nose bob' was born.

Cosmetic surgery had come of age as a way of reconstructing the shattered faces of soldiers injured in the World Wars of the twentieth century. With the new surgical refinements achieved, it was then realized that the same procedures could be applied for purely aesthetic reasons, where someone was unhappy with the face that nature had given them. Reducing the size of a woman's nose became by far the most popular operation.

The technical term for the nose bob, or nose job, is *rhinoplasty* – Greek for 'nose-moulding'. The surgery is performed inside the nose so that there are no external scars. A typical case involves the removal of a bony hump that is making the nose too prominent and hooked. A special saw removes this hump of bone and the nasal profile is dramatically reduced. Less common modifications are the reduction of an unusually bulbous nose-end, nostril trimming and the elevation of drooping nose-tips.

As often happens with new developments in body 'improvements', some of the earliest customers for nose bobs were the stars of show business. In 1923 the then famous American theatrical performer Fanny Brice summoned a famous cosmetic surgeon to her apartment at the Ritz where he performed rhinoplasty to reduce her prominent nose to more petite dimensions. Her employer was horrified, exclaiming that she had had a inimitable 'million-dollar nose', and Dorothy Parker, famous for her caustic comments about the celebrities of the day, acidly remarked that Brice (who was Jewish) had 'cut off her nose to spite her race'. The actress vigorously defended her action. Later, in the 1960s, when Barbra Streisand played the role of Fanny Brice in the film *Funny Girl*, she bravely refused to modify her impressive nose, and the original Brice nose bob incident was omitted from the script of the film biography.

Streisand, her resolve strengthened by her powerful personality, was, however, an exceptional case. In the second half of the twentieth century nose bobs became increasingly popular in the Western world, as actresses, models and, indeed, women from all walks of life had their large noses trimmed down to more modest proportions. By the twenty-first century, the number of noses that had been surgically modified could be counted in their hundreds of thousands.

Even in the natural homelands of the larger noses, the custom began to spread. Cosmetic surgeons in Israel, for example, have been kept busy trying to meet the demands for more and more nose reductions. In addition to the local population, the socially aware young women of Egypt, Jordan, Saudi Arabia and the Gulf States have all been flocking to Israeli clinics to have this done.

The procedure became popular even in the most unexpected places. In the strict Islamic regime imposed in Iran, where women must cover their hair in public and expose only their faces, or parts of their faces, it is astonishing to discover that rhinoplastic surgery is flourishing. At the start of the twenty-first century it was reported that nose reductions had become such an obsession with smart young Iranian women that more than 100 rhinoplastic surgeons were performing 35,000 nose operations per year. A Tehran teenager commented, 'It is such a trend that even if people don't get a nose job, they will wear tape for the attention it brings.' Their excuse is that, according to Islamic teaching 'God loves beautiful people', but the real reason, of course, is that, with almost all the rest of their anatomy concealed by the Islamic dress code, the nose has become a main focus of attention.

In some regions of tropical Africa, a different kind of operation was gaining in popularity. With this, the unduly flat, wide noses of the local women are made narrower and given a firmer nose-bridge. This is the nasal equivalent of straightening crinkly hair and is an attempt by fashionable young African women to look more European. A similar trend has recently been reported from the Far East. In Vietnam and China, noses are now also being Westernized in large numbers.

As a site for jewellery the female nose has not been as popular as the ears, the neck, the wrist or the fingers. In some tribal societies the nasal septum has sometimes been perforated for the attachment of hanging ornaments, but this has never become widespread. Nostril piercing has a longer history, having started in the Middle East about 4000 years ago. It is still a common practice among the nomadic Berber and Bedouin peoples of North Africa and the Middle East, where the husband gives his bride a gold nose ring when they are married. The size of the ring indicates the wealth of the family, and, if a divorce occurs later, the rejected wife can use the gold in her nose to provide her with security.

The nose-piercing tradition was carried from the Middle East to India during the Mogul period in the seventeenth century, where it became the custom to pierce the left nostril. The left one was chosen because, in local superstition, this side was connected with female reproduction and childbirth. It was believed that, if a nose stud was worn (and sometimes connected to the left ear by a gold chain), it would render childbirth less painful.

In the 1960s it became popular for Western hippies to trek to Asia to 'find themselves' and, when they saw the pierced noses of local women, many decided to adopt this exotic form of body mutilation. Back in Britain, it was taken up, in a crude form, by the punks of the 1970s, but was still looked upon as an exotic form of body piercing. Later, towards the end of the twentieth century, perhaps under the spreading influence of Bollywood films, small, jewelled nose studs became increasingly popular. In many places there were angry reactions from employers and sacking of employees for wearing this new type of female decoration, but eventually the custom became so accepted that it lost some of its 'rebel appeal' and now, in the twenty-first century, it is already on the wane.

Nose contacts have always been rare in social contexts. Public, interpersonal contacts with the nose in Europe have usually been brutish and nasty. The best a nose could hope for would be a tweak or a punch. The worst was a particularly savage form of punishment – the nose slit – in which a knife was inserted into the nostrils and the nose then sliced open. This was introduced in the ninth century as a way of dealing with those who failed to pay their taxes.

Although today tax collectors have put away their knives, we still have a relic of their early methods, in the popular saying 'to pay through the nose'.

Only between lovers in private was the nose offered more gentle touches in the Western world. During lovemaking, nuzzling, nose-pressing and nose-kissing have always been common, but they have never developed outside the context of sexual intimacy. Among Pacific islanders they occur in both sexual and non-sexual situations. Here is Malinowski's translation of a Trobriand man's description of his lovemaking: '. . . I embrace her, I hug her with my whole body, I rub noses with her. We suck each other's lower lip, so that we are stirred by passion. We suck each other's tongues, we bite each other's noses, we bite each other's chins, we bite cheeks and caress the armpit and the groin . . .'

In purely social contexts the peoples of the Pacific region employed nose-to-nose contact in much the same way that we would use a social kiss. Their action is usually referred to as 'nose-rubbing', but this is an error. The rubbing movements are normally reserved for the erotic encounters of the kind described by Malinowski. In public the action is little more than nose-tip touching or pressing. It is based on the concept of mutual smelling with each nose inhaling the fragrance of the other's body.

As a formal greeting, nose touching is sometimes subjected to rigid rules of status. In one culture, the Tikopia, found in the Solomon Islands in the South Pacific, there is a whole range of parts of the body that may or may not be touched with the greeter's nose. Nose-to-nose or nose-to-cheek contact is only allowed between social equals. When a junior meets a senior the contact must be nose-to-wrist. When a follower greets a great chief, it must be nose-to-knee.

Nose greetings are on the decline today. The more cosmopolitan way of life, increased travel and mixing of cultures, more and more tourism and international trade, have all contributed to a greater uniformity of greeting gesture, with the ubiquitous handshake spreading to cover almost the entire globe. Nowadays, when high-ranking Maoris meet they combine a vigorous handshake with a fleeting nose-touch – the new edging out the old.

7. The Cheeks

Since ancient times the soft, smooth, female cheeks have been thought of as a focal point of human beauty, innocence and modesty. This is partly because the exaggerated roundness of the cheeks of a baby – a feature unique to humans – acts as a powerful infantile stimulus releasing strong feelings of parental love. This early connection between smooth cheeks and intense love leaves a residue in our adult relationships. In our more tender moments we reach out to touch, kiss, stroke or gently pinch the cheeks of a loved one, homing in on this part of the anatomy because of its associations with the pure love between parent and child. Just as the young mother presses her infant's cheek gently against her own, so do lovers dance cheek to cheek and old friends kiss and embrace cheek to cheek. Symbolically the cheek is the gentlest part of the whole female body.

The cheek is also the region most likely to expose the true emotions of its owners. For it is here that emotional changes of colour are most conspicuously displayed. The blush of shame or sexual embarrassment begins at the very centre of the cheeks – at two small points that turn a deep red – to be quickly followed by the rest of the surface of the skin of the cheek and then, if the blush intensifies still further, by other areas of skin such as the neck, the nose, the earlobes and the upper chest. Mark Twain once exclaimed that 'Man is the only animal that blushes. Or needs to . . .' – as if it were the terrible misdeeds of human beings that caused their cheeks to flame red with shame. But this is not the context in which the blush is observed. The typical blusher is young, self-conscious, socially rather shy and usually has nothing much to be ashamed

71

about except personal inexperience and unwanted innocence in an atmosphere of sophisticated knowingness.

The fact that blushing crops up repeatedly in erotic situations makes it look rather like a special sexual display of virginal innocence. The 'blushing bride' is a popular cliché of marriage ceremonies, the blush here being the result of a self-consciousness about the thought that everyone present is privately contemplating the young woman's imminent loss of virginity. Because blushing is (or was, before modern sexual education led to a greater openness and frankness on the subject) closely linked to courtship contexts and flirtation moments of very young adults, it has come to be linked with sex appeal. The woman who does not blush is either unaware of her own sexuality or is brazen about it. The woman who blushes when a sexual remark is made is obviously aware of her own sexuality but is still unsophisticated. Therefore it could be argued that blushing is basically a human colour-signal denoting virginity. In this connection it is significant that the young women being offered at ancient slave markets for use in harems fetched much higher prices if they blushed when being paraded before potential buyers.

The female cheeks also act as indicators of anger when they flush bright red. This is a different pattern of reddening, a general diffusing of the colour rather than a spreading from the centre of the cheek. The mood of the angry woman is one of inhibited attack. She may issue dire threats, but the red skin indicates that the mood is a frustrated one. The cheeks of the truly aggressive woman turn very pale, as the blood is drained away from the skin, ready for immediate action. This is the face of the woman who is really likely to leap into the attack at any moment. Similarly, if she is intensely frightened the emotion of fear will also make the cheeks blanch, ready for the action of fleeing or of striking out if cornered.

In modern times, the bronzed cheeks of the (Caucasian) sunbather offer the status signal of a woman who has been able to take time off to lie in the sun on some holiday beach. This is a comparatively recent development. In earlier centuries no high-status young female or 'young lady of fashion', as she would have been called then – would have been seen dead with suntanned skin. In those days tanned skin meant only one thing – peasant toil in the fields. Upper-class young ladies would have looked upon a tanned skin as utterly repugnant and would even have taken special measures to avoid the slightest tinge from a stroll in the park, by wearing a sun-shading hat or bonnet, or carrying a parasol.

At certain periods in history this anti-sun attitude led to the whitening of cheeks with the aid of make-up. In more extreme cases, women would go to the length of bleeding themselves to make their cheeks paler. Neither of these practices was without its dangers. White make-up in the sixteenth century was especially hazardous because it contained lead oxide. Repeated use of this type of cheek-paint resulted in the cumulative storage of poison in the body that eventually caused muscle paralysis and sometimes even death.

At other times, when it was thought that rosy cheeks as distinct from bronzed ones were a sign of good health and natural vigour, the centres of the cheeks were painted with rouge. If rouge was not worn, young ladies could be found pinching their cheeks outside an important social gathering to bring the blood to them.

'Blushers' are still popular in female cosmetics to this day, although they tend to come and go year by year as the fashion houses struggle to keep novelty alive for commercial purposes. This form of make-up carries not only the pseudo-healthy signals but also a suspicion of the teenage blush of innocence as well, giving it a double advantage in sexual contexts.

In the twenty-first century, following a medical campaign in which over-enthusiastic sunbathing has been directly linked with skin cancer, the bronzed cheek has once again fallen out of favour. Many young women now avoid grilling themselves on holiday beaches and use heavy sunblock creams, or avoid the sun altogether. The pale cheek is once against a popular symbol – this time of the responsibly health-conscious. Some women, however, still refuse to abandon their sun-worshipping ways, and holiday crowds are now split between the cautious paleskins and the risk-taking bronzers. It remains to be seen which group will finally prevail.

If suntanned skin can sometimes cause medical problems, they are minor compared with the impact of a cheek-cream that was sold under the name of 'Aqua Toffana', or 'Manna of St. Nicholas of Bar' in seventeenth-century Italy. A certain Signora Giulia Toffana offered this very special face-treatment for sale to fashionable ladies and it proved to be particularly popular among wives who wished to dispose of their husbands. Sold as a cream or powder, it was a highly poisonous mixture containing arsenic and other lethal ingredients. Signora Toffana always insisted on a special visit from each of her clients, so that she could instruct them in its proper use. She explained to them that they must never ingest the make-up and must apply it to their cheeks when their husbands were about to engage them in amorous contact. This ensured that their husbands' mouths, pressed against their cheeks, would take in enough of the make-up to kill them. Afterwards, the excuse was always 'death from sexual excess' and the ruse worked well for many years.

Toffana was responsible for over 600 deaths and the creation of the same number of wealthy widows, making her the greatest poisoner of all time. Her crimes were uncovered in 1709, when she was arrested, tortured and strangled in prison.

Apart from colour, the shape of the cheeks is also important. A dimpled cheek has always been considered attractive in Europe because the dimples are said to be the mark made by the impression of God's finger. Dimples do not seem to be common today and they have probably always been rather rare, which may account for the unusual amount of folklore and superstition attached to them. There are many old rhymes and sayings about them, such as 'A dimple in your cheek / Many hearts you will seek . . .' and 'If you have a dimple in your cheek you will never commit murder.'

Among the early Greeks the shape of the cheeks was also important as a standard of beauty, and the Greeks had a special gesture for it: the Cheek Stroke. This consisted of placing the thumb and forefinger of one hand high on the cheeks, the thumb on one cheekbone and the forefinger on the other. From this starting point the hand is stroked gently down the cheeks towards the chin and off. During this movement the thumb and forefinger are brought gradually closer towards one another, suggesting a tapering shape to the face. It was this egg-shaped face that the Greeks considered as the ideal of feminine beauty. Modern Greeks still interpret the gesture in this way.

The word 'cheeky', meaning impudent, stems from the gesture

known as 'tongue-in-cheek' in which the gesturer signals disbelief by pressing her tongue hard into one cheek so that the shape of the cheek is distorted. This originates from the idea that the only way the gesturer can prevent herself from saying something critical is to press it hard into her cheek, to stop her uttering the words that are 'on the tip of her tongue'. To show 'cheek' in this way was considered rude, especially if done by children, and in the early Victorian period the terms 'cheek' and 'cheeky' entered the British language.

Another gesture, largely confined to Italy, is the Cheek Screw. The forefinger is pressed into the cheek and twisted round as if screwing something into the flesh. It is known by almost everyone, from Turin in the north to Sicily and Sardinia in the south of the country. It always has the same meaning: 'Good!' In origin it is a compliment to the chef signifying that the pasta is *'al dente'* or 'on the tooth'. In other words, the food is cooked to just the right consistency, as suggested by the forefinger pointing at the teeth inside the cheek. But as time passed it became used in a wider and wider context to include anything good. When used of a young woman, it is roughly equivalent to the English expression 'Very tasty!'

Placing the palms together and then resting one cheek on the back of one hand is a widespread sign meaning 'I am sleepy', based on the fact that the moment which most typifies the activity of sleeping is when the cheek hits the pillow. It is interesting that when people are tired or bored but have to remain sitting at a desk or table they are most likely to adopt a resting posture in which one hand supports a cheek as if propping up a heavy head. When lecturers or teachers see this posture they should realize that they are in trouble. A more obvious sign of boredom is the Cheek Crease, in which one mouth-corner pulls back hard to bunch up the flesh of the cheek. This also signifies disbelief and is essentially a gesture of heavy sarcasm.

In some parts of the Mediterranean pinching your own cheek is a signal that something is excellent or delicious. Almost everywhere this same action, but performed on someone else's cheek, is a sign of affection. It has been used in this way for over 2,000 years, having been popular in ancient Rome. It is normally used by adults towards children (who frequently hate it), but may also be performed in a joking way between adults.

The Cheek Pat with the palm of the hand is employed as a slightly less irritating alternative, but this too can become annoying when it is performed with too much vigour. In cases of false affection this patting action may easily become magnified into a near slap, leaving the victims in an awkward state of knowing that they have been insulted but unable to do anything about it because the action is so close to the friendly gesture.

The Cheek Slap itself has a long tradition. It was the classic action of a lady responding to the unwelcome attentions of a male. In essence the Cheek Slap is a 'display-blow' – a blow that makes a great deal of noise but causes so little physical damage that it does not provoke an immediate defensive or aggressive action on the part of the victim. Although it instantly pulls the recipient up short, its significance sinks in later.

At the other end of the emotional scale are the Cheek Kiss, the Cheek Touch and the gentle Cheek Stroke. The Cheek Kiss is a reciprocal action suitable only for two people of equal status. It is a diverted, low-powered Mouth Kiss and has become widespread in many countries as part of the ritual greetings and farewells of social gatherings. Where lipstick is worn it is often more of a cheek-to-cheek press combined with a kissing noise but without lip-to-cheek contact. There are considerable sub-cultural variations in its frequency. In theatrical circles and in the more flamboyant social spheres it is almost over-used, whereas in 'lower income' areas it is often extremely rare except between close family. This difference varies as one moves from country to country. In parts of Eastern Europe, for example, the original mouth-to-mouth greeting kiss remains common and is not diverted on to the cheek.

Mutilations of the cheek region have not been particularly popular because of the need for facial mobility in so many situations. In ancient times there was, however, a custom among female mourners of scratching their cheeks and making them bleed as the most obvious way of displaying their agony. John Bulwer reports that this led to a 'keep-cheeks-smooth' law being passed: 'The Roman dames of old were wont to tear and scratch their cheeks in grief

. . . insomuch, as the Senate taking notice thereof, made an edict against it, commanding that no women should in time to come, rent or scratch the cheeks, in grief or sorrow, because the cheeks are the seat of modesty and shame.'

Tribal decorations of the cheeks include a variety of face-paintings, tattooings, incisings and hole-borings. Apart from the simple powder-and-rouge routines mentioned earlier, the Western world is comparatively free of these facial adornments, although there was a brief resurgence of them in the 1970s with the punk rock movement in London, when women could be seen with a safety pin inserted in the flesh of their cheeks, usually close to the mouth. These savage mutilations of the early Punks gradually softened and fake safety pins were eventually put on sale, giving the impression of being skewered through the flesh without actually harming it.

The only other form of cheek decoration that deserves a special mention is the 'beauty spot' that became the height of fashion in the seventeenth and eighteenth centuries. This began as a cover-up for small blemishes, but soon took on a cosmetic life of its own. A story was put about that Venus was born with a natural beauty spot on her cheek and that any lady of fashion who chose to emulate her could therefore only gain in beauty. This provided the excuse for covering a mole, wart or pockmark with a small, circular black patch, or disguising it with a black make-up pencil.

This form of cheek decoration became so popular that women with perfectly smooth skins joined in and adopted face patches and beauty spots as a purely decorative device. Eventually they became so essential in court circles that, at one stage, it was said that 'all fashionable ladies should wear them all the time, unless they were in mourning.' At the end of the seventeenth century, a sharp-tongued Frenchman visiting London was moved to comment, 'In England the young the old and the ugly are all bepatched until they are bedridden. I have often counted fifteen patches or more upon the swarthy, wrinkled face of a hag three-score and ten upwards . . .'

By the early eighteenth century, the fashion had became so intricate that the position of the beauty spots even developed political significance, with (right wing) Whig ladies decorating the right cheek and (then left wing) Tory ladies decorating the left cheek. The beauty marks themselves ceased to be mere spots and were elaborated into stars, crescents, crowns, lozenges and hearts. These excesses were soon to disappear, but an occasional, single beauty spot could still be seen from time to time – a simple survivor from a complex past.

In modern times, with a few notable exceptions, this fashion for beauty spots has waned and today's blemished female cheek is offered a different kind of treatment. Because a smooth, unblemished cheek suggests that its owner is both youthful and healthy, it is doubly important for an attractive young woman to conceal any acne scars, blotches, roughness, wrinkle lines, or other forms of skin damage. If ordinary make-up cannot mask the problem, something more severe is needed. To this end, several new procedures have been introduced by cosmetic surgeons.

One is skin abrasion, or, to give it its technical term 'micro-dermabrasion'. In this, the cheeks are virtually sandblasted into smoothness. A stream of sand-like aluminium oxide crystals is aimed at the cheek, removing the outer layers of skin. After healing, the skin is much smoother, if the treatment is successful.

Another procedure is chemical skin-peeling. A thin layer of a special peel gel is applied to the cheeks and then, after about five minutes, carefully wiped away. This acidic gel removes the damaged outer layers of the skin.

A third method employs a high-tech combination of ultrasound, micro-current, skin-vacuum and laser treatment.

In all three cases the procedures usually have to be repeated a number of times and the results are not always perfect, but new advances in this kind of cheek-improvement are being made all the time and the day will soon come when any woman will be able to buy perfectly smooth cheeks – at a price.

8. THE LIPS

There is something very strange about human lips. Uniquely in the animal world, they are turned inside out. Most people do not realize this because our lips are taken for granted and we never bother to compare them with the lips of our primate relatives, the apes and monkeys. But if we looked closely at the mouth of a chimpanzee or a gorilla we would soon see that, there, the soft, fleshy, shiny surface that we see when we look at any human mouth is hidden from view.

Why do humans have these everted lips? Once again, the answer is connected with our evolution as childlike adults. As our adult anatomy and behaviour became progressively more infantile, we retained more and more 'babyish' features, and our visible, fleshy lips are part of this trend. And because the human female is slightly more advanced anatomically – that is to say, more juvenile – than the male in this respect, it follows that her lips are, on average, more conspicuous and protuberant. As a result, they have become the focus of a great deal of attention.

But first, where can we track down the origins of these super-lips? The answer lies, not in a human baby, or even a chimpanzee baby, but in a tiny chimp embryo. When the ape foetus is only sixteen weeks old, it has a typically humanoid mouth, with big, puffy lips. Two months later, when it is about twenty-six weeks old, they have already gone. They have shrunk back to the thin-lipped type that will remain with the ape for the whole of the rest of its life. So, to be accurate, human lips are not merely juvenile, they are embryonic.

Unlike the chimp baby, the human baby clings on to the early foetal design and at birth it gurgles at its mother with a pair of pouting *outrolled* lips that are soon pressed joyously on her short nipples, where they squeeze the milk from her very rounded breasts. The little chimp is instead clamping its thin-lipped, muscular mouth on its mother's long teat and working the milk from her like a farmer milking a cow.

So the uniquely everted human lips are well suited to their first task as milking devices on the equally unique human female breasts. They provide an airtight seal on the rounded surface they encounter. But the story does not end there. If it did, the baby-lips would turn in on themselves as the infant was weaned on to solid foods, and then follow the typical, thin-lipped, primate pattern into adulthood. In the adult human male they do, indeed, become slightly tighter and thinner and, in a primitive condition, eventually disappear from view beneath masculine facial whiskers.

The typical human female, however, continues to display a pair of full, soft lips for the rest of her adult life – or at least until she reaches very old age, when she too joins the thin-lipped. As a young adult with sex on her mind, she now starts to treat her lips as a new kind of signal – a strongly sexual one. She moistens them, pouts them, blows kisses with them and decorates them. Even before she has placed them on her first lover's mouth, they will have played a major part in her feminine presentation.

What makes the lips visually so sexual? The answer is that in their shape, their texture and their colour, they are mimics of the female's other lips, the ones so intensely sexual that, even today, they are spoken of in a classical language – *labia* being Latin for *lips*.

When the human female becomes sexually aroused, her labia become reddened and tumescent. At the same time, on her face, her lips also become swollen, redder and more sensitive. These changes occur in unison, as part of the physiological upheaval that accompanies extreme sexual arousal. One of the key factors in this process is a shifting of blood from the deeper organs to the surface. The skin of the sexually active individual glows and shines as the tiny blood vessels become distended with an increased blood supply. This extra blood arrives more quickly than it can depart and, as a result,

80

the body surface becomes more and more sensitive to touch. This is especially true of the lips. The distended blood vessels also make the lips and the labia more conspicuous, their increased redness contrasting more and more with the surrounding flesh.

Intuitively, the females of early societies began to exploit this lips/labia mimicry. Prostitutes in ancient Egypt used red ochre to heighten the colour of their lips. There is a papyrus drawing dating from 1150 BC that shows a scene in a Theban brothel where a semi-naked young woman is holding up a mirror and applying lip colour with a long stick. A balding client with a large erection is depicted in the act of moving his hand towards her genitals. The connection between reddened female lips and erotic activity is therefore more than 3,000 years old.

The use of some kind of lip colouring is even older than that, for evidence of its existence has been dated back to the city of Ur, in what is now southern Iraq, four and a half thousand years ago, where a great queen, Queen Puabi, was buried with an ample supply of make-up for use in the afterlife. Her cosmetic paints were kept in large cockleshells, or in imitation shells made of silver or of gold. There was red paint for her lips and white, green and black, presumably for her eyes.

The manufacture of the earliest lip colourings was achieved by grinding red ochre to a fine powder, using a small pestle and mortar, and then mixing this powder with animal fats. Later, in the fourth century BC, the ancient Greeks were more experimental and are reported to have added certain plant dyes, human saliva, sheep sweat and even crocodile dung to the mixture. In the second century AD, women in Palestine had advanced to the stage where they even had a choice of colour – bright orange-red or deep raspberry-purple.

Since those days, the artificial reddening of the female lips has been a popular and recurrent theme of feminine display, although it has sometimes fallen foul of puritanical authorities. In cultures where restrictive regimes have imposed themselves and have sought to suppress sexual pleasures, lips have remained unadorned. In extreme cases, even the unadorned lips have been considered too exciting to be seen in public and unfortunate women have been forced to conceal them beneath veils of cloth.

The hidden lips of veiled women are generally thought of as a requirement of the Islamic faith, but this is not the case. It is true that this concealment occurs widely in Moslem countries, but it has nothing to do with the teachings of Mohammed. It has been imposed on women by their male-dominated society. It is not a religious statement, but a sexist one, in which women are callously treated as male property.

The Christian Church has had an ambivalent attitude to red female lips. At times it has been open-minded, but there have also been periods of ruthless suppression when artificially coloured lips have been viewed as evil and as a vile challenge to God's handi-work – the natural human body. One seventeenth-century clergyman condemned painted lips as 'the badge of a harlot' – an ensnarement that kindled a fire of lust in the hearts of men unfortunate enough to set eyes on them.

Politicians have generally kept out of such matters, but at one point in eighteenth-century England, they felt obliged to pass a law banning the use of lip colouring because certain anxious males felt that they might be falsely lured into wedlock by the sight of red-lipped women. This nonsensical state of affairs created something of a problem for the young ladies of the day. Their solution was to suck on grenadine sticks or pinch their lips just before making their entrance at a social gathering.

Despite repeated suppressions by Church and State, lip cosmetics refused to disappear and throughout history kept on bouncing back in one fashionable style or another. In a *Lady's Magazine* of the late 1820s, it is clear that a special lip-shape has been adopted – the cupid's bow. In this, the lips are enhanced vertically instead of side-ways. The lower lip is deeper but not wider, and the upper lip shows a marked dip just below the nose. This gives the female mouths a babyish look, and transmits the appealing signal to the gallant males of the day that these fair maidens need their protection.

Moving on to modern times, the application of lipstick has come to support a major industry. This industry grew gradually throughout the twentieth century. By the end of the Victorian era bright red lips had become largely confined to the infamous Houses of Pleasure, driven there by the prudery and double standards of the day. Countless male clients were excited by their blatantly inviting colour, before returning home to their pale-lipped wives.

Then, during the First World War, lipstick began its slow climb up the social ladder, spreading from the brothels to the theatres and from the theatres to the more daring members of Bohemian society. After the war, in the roaring twenties, the red lips spread further still, to smart young things on the dance floor. In the twenties and thirties, lipstick was also taken up by the great stars of the rapidly spreading cinema screen, and had soon become the social norm.

One of the earliest film stars, Clara Bow, the original It Girl, re-introduced the cupid's bow lips, but in a bolder, more vivid form. With her baby-face, she was appropriately christened the 'Hottest Jazz Baby in Films' and in 1925 starred in a film entitled *My Lady's Lips*. In the 1930s, more dominant females appeared on the scene and changed the lipstick style to an imposing, red gash. After this, the babyish cupid's bow disappeared.

By the outbreak of the Second World War, among the young at least, applying a bright red colouring to the lips was seen as the patriotic thing to do – to cheer up the brave fighting boys. Recruitment posters showed vividly red female lips that were clearly intended to offer the promise of female support to anyone who would defend their country.

In 1945, at the close of the Second World War, a period of austerity began. Peace may have returned, but lipsticks were now a trivial luxury and came in only a few different colours – always bright reds of one shade or another. The use of any other colour was unheard of. In the 1950s this was all to change. In France and Italy at about this time, cosmeticians introduced titanium white into their lipsticks to produce paler colours and in so doing started a dramatic increase in the range of colours and shades. The fashion magazines of the day held great influence and had the power to introduce a new colour each year – a colour that became the craze for a whole season, then vanished, to be replaced by 'the latest thing'.

In the 1960s, with the arrival of the contraceptive pill for women, open sexuality enjoyed a field day and women began to express themselves more vigorously as individuals. Instead of one dominant lipstick colour, there was a whole range of startling, rebellious shades and hues from which to choose, including many very pale shades.

With the arrival of feminism in the 1970s this quickly changed. Painting one's lips was, for a while, looked upon as pandering to male desire, and a new kind of female-dominated puritanism came to the fore. Feminist lips went unadorned. At the same time, women were protesting widely against the Vietnam War, and, if they were outside the feminist movement, sometimes adopted dark, forbidding lip colours such as blue, purple or even gothic black.

Once the Vietnam War was over and young females had gained a greater social equality, the more severe forms of battledress were discarded and successful women now felt free to look like women again. In the 1980s and 1990s bright red lipstick returned once more.

By the start of the twenty-first century, young females had begun to express their sexual desires more honestly than ever before and, with this greater sexual confidence and openness came increasingly erotic promotions of lipstick products. There were three basic strategies – intensely red lips, brighter than ever before; natural-coloured lips rendered shiny by lip gloss; or a combination of the two – very red and very shiny. Individuality was now the key. Women were no longer slaves to one, over-riding fashion rule. Each could make up her own mind. At a pop concert, one female performer could appear with vivid, blood red lips and the next come on stage with glossy pink lips – or occasionally no lipstick at all.

Advertisers became increasingly obsessed with descriptions of ultra-shiny lips, juicy lips, luscious lips, wet-look lips and mouth-watering shades. Their accompanying colour photographs displayed female lips of such glistening wetness that it is impossible to avoid the underlying biological message, namely that if intense sexual arousal leads to genital secretions, the new lipsticks must somehow suggest this physiological change. These lipstick manufacturers did not create an enhanced mouth, they created a pair of super-labia. The message is now clear for all to see – women are actively displaying their enjoyment of sex and they do not care who knows it.

Impressive as they are, all these Western techniques of female lip enhancement pale into insignificance alongside the lip-stretching mutilations of some tribal societies. In the Surma people of south-west Ethiopia the adult females are known as 'plate-women'. When they are in their early twenties, six months before they are to be married, either the lower lip or the upper lip is cut away from the rest of the face and a small plate, called a labret, is inserted in the incised hole. This stretches the lip out from the face in a rubbery ring of flesh. As soon as they can bear to do so, the young women remove the small plate and replace it with a slightly larger one, then a larger one still, until they are displaying a lip labret almost the size of a dinner plate. In earlier days these plates were wedge-shaped and carved out of wood, but more recently it has become the fashion to make them circular and of baked clay. When the women are alone, or eating, sleeping or in the company of other women, they are allowed to take the labrets out. When they do this, their stretched, incised lips hang down loosely from their faces. When men are present, however, the plates must always be worn, and act as a badge of status, the size of the plate denoting the worth of the woman. The dimensions of the largest plate that a particular young woman can tolerate will be used as a measure of her beauty and also to determine how many cattle she is worth when her hand is offered in marriage.

This bizarre form of lip enlargement has existed in quite a number of distinct African tribes, not only the Surma, but also the Makonde of Kenya, the Lobi of Ghana, and the Sara-Kaba and the Ubangi of the Congo basin. Amazingly, for such an extreme form of body decoration, it was also discovered by early explorers in a completely different part of the world – on the west coast of Canada, where Tlingit Indian women in British Columbia displayed large lip-discs. Once again, it was the women with the largest discs who enjoyed the highest status.

The precise technique varies from tribe to tribe. Some stretch only one lip, some both lips, while others fit wooden pegs into holes both above and below the lips. In all cases the effect is to enlarge the lip region and make it the focus of attention. In the case of the Ubangi, it has been suggested that the tribal chiefs instigated the procedure to deter Arab slave traders, who found the plate-wearers unappealing and went elsewhere in their search for slave-girls. Although this story has been widely reported, there appears to be little to support it. What is much more likely is that the Ubangi, like the other lip-plate tribes, saw the stretched lips of their women as a mark of beauty, and that the distaste of the slave traders was merely an added bonus.

Other tribes employ different techniques. The Shilluks of the Sudan prefer blue-stained lips on their women. The Ainu of Japan like their females to have tattooed lips. The tattooing starts in child-hood and by the time the girls are adult the blue-black lip tattoo has spread right out from the mouth until it reaches almost from ear to ear. In some Philippine tribes a chewing gum made of betel nut is used to incarnadine (redden) the lips.

When early explorers first set eyes on these extravagant lip fashions, they sometimes found it hard to believe that they were self-inflicted, assuming that the women of the tribes had evolved such lips: '. . . they are naturally born with their lower lip of that greatness, it turns again and covers the great part of their bosom, and remains with that rawness on the side that hangs down, that through the occasion of the sun's extreme heat, it is still subject to putrefaction; so as they have no means to preserve themselves, but by continually casting salt upon it'. This account was written by John Bulwer as early as 1654, in one of the first anthropology books ever published, and it clearly did not occur to him that the health problems involved with these huge lips resulted from the surgical creation of raw surfaces, where the lips had been cut away from the mouth, for the fitting of the large discs.

Cosmetic lip surgery, once so common in African tribal societies, was nowhere to be seen in urban society for many centuries, but recently re-surfaced in a new form in California. Hollywood actresses, aware of the sex appeal of thick, luscious lips, began having their natural beauty enhanced by a variety of surgical proce-dures. Without going into technical detail, the main points of this type of cosmetic surgery can be summarized as follows (although it must be emphasized that new procedures are being introduced all the time):

The least drastic procedure is a series of syringe injections, of either collagen or hylaform gel, at various points along the top lip, then along the bottom lip. The effect lasts for about three to six months and an actress will sometimes employ this type of lip enhancement for a specific role in a particular movie.

More permanent enlargement requires a procedure involving the surgical excavation of a small tunnel in the patient's lips, reaching from one mouth-corner to the other. Into this hollowed-out space, additional material is inserted, to swell up the lip tissue. The substance added may take the form of strands of synthetic material; or alloderm, which is freeze-dried skin; or the patient's own fat, taken from her buttocks, purified and then re-injected into her lips.

Finally, there is the most extreme form of enhancement – lip surgery. This is a permanent reshaping of the lips by the surgeon's knife and has to be carried out in an operating theatre. The procedure takes about an hour and has the disadvantage that it leaves scars. Although these scars are invisible, being positioned inside the mouth, they can still be felt.

In each of these cases, the augmentation of the lips can be designed to increase one of two qualities: the general fullness of the lips, or their forward projection. Which of these two features is enhanced will depend on the precise placing of the added substances. Sometimes the fuller-lips strategy has the odd effect of eliminating the cupid's bow shaping of the top line of the upper lip. Instead of dipping in the middle, the top line curves smoothly across beneath the nose, creating a slightly artificial look.

Another risk with all these surgical procedures is that, after the lips have been modified, they no longer suit the face that surrounds them. There are already certain actresses whose 'bee-sting lips' look so impressive that they overpower the rest of their facial features (and in some cases have been cruelly referred to as a 'trout pout'). Critics have voiced the opinion that, unless a young woman is excessively thin-lipped, it is wise to think twice about undergoing this type of surgery. But such comments are going largely unheard, and the twenty-first century has seen a rapid spread of this type of cosmetic procedure, from its beginnings in California, across the United States, to Europe and beyond. It is certainly true that, if the lip specialist does his or her job well, and avoids the pitfalls mentioned here, a female face can suddenly become much sexier, such is the erotic impact of the human female lip-shape.

Up to this point, the lips have only been considered as visual signals, but, of course, they are not there merely to be seen. In a recent survey of the ten most important contact points on a woman's body (for a man to touch during foreplay), the number one female erogenous zone was named as the lips. Not the breasts or the genitals, but the lips. It is true that, in the later stages of lovemaking, stimulation of the clitoris is most likely to lead to orgasm, but during the earlier phase of sexual foreplay, it is contact with the lips that is the biggest arousal factor, according to the women interviewed for the survey. This may explain why, traditionally, prostitutes say 'no kissing', despite the fact that they permit all kinds of genital contact. When asked for a reason for their kissing taboo they are reported as replying that it is not because mouth-to-mouth kissing in unhygienic, but because it is 'too personal', a comment which says a great deal about the significance of the female lips. It may also explain why, in some countries such as Japan, there is a taboo on kissing in public.

Erotic mouth kissing has an intriguing origin. When lovers bring their open lips together and start probing the interior of one another's mouths with their tongues (the so-called deep kiss, French kiss or soul kiss) they are performing an action that harks back to primeval times. Before there was convenient 'baby food' available, tribal women used to wean their infants off breast milk and on to solid foods by pre-chewing the food in their mouths until it was soft and semi-liquid. They then placed their open mouths over those of their babies and, using their tongues, passed it over into their babies' mouths. As the infants grew used to this, they began to probe for the soft food with their own tongues, as soon as mouth-to-mouth contact was made. In this way, the action of tongue-probing became indelibly associated with a loving act.

From this ancient beginning grew the deep kissing of loving adults. We have forgotten how this came about because today it is extremely rare to find surviving examples of the primeval food-kissing. It does still occur in some remote tribal societies but is now unknown and long forgotten everywhere else.

It is worth pointing out that, because of the very high tactile sensitivity of the female lips, their application to different parts of the male body during sexual foreplay and lovemaking is less altruistic than it may seem. According to the classic study of human female sexuality by Kinsey and his colleagues, published half a century ago, some women may even be able to arouse themselves to the level of orgasm during prolonged bouts of mouth-to-mouth deep kissing, and this can occur despite the absence of any kind of genital contact.

A few women may also be able to achieve orgasm when applying their lips to the male phallus. The female may appear to be involved purely in servicing the male and arousing him, but so refined are the nerve-endings in the mucous membranes of the everted female lips, that every touch they make on a loved one's body also sends back powerful stimuli to their owner. In this respect, as in so many others, the human female is the most highly developed of all the primates.

Oral–genital contacts – which we now know are not the modern inventions of 'decadent' Western society but have played a major role in the sexual activities of many cultures for thousands of years – are strongly related to infantile oral pleasures at the breast. When a young lover kisses the penis of her partner her mouth movements are strongly reminiscent of those she enjoyed when, as a baby, she was suckled by her mother. The impression made by that early oral stage of life stays with her in some shape or form throughout much of her adulthood.

It should be added that the Freudian view of adult oral pleasures is that they reflect infantile *deprivation*. The suggestion is that infants denied the oral rewards normally provided by mothers will spend the rest of their lives trying to compensate for the loss. In extreme cases this may well be so, but what Freud overlooked is that the pleasures experienced at any stage of life are liable to establish patterns of behaviour for the future. An individual who as a baby enjoyed sucking at the breast, as most do, is hardly likely to forgo the chance of enjoying adult ways of recapturing such a pleasure – simply because there was *no* infantile deprivation. Freud's negative attitude towards adults who enjoy kissing, smoking, eating sweet foods and sipping warm sweet drinks is perhaps not hard to understand because his own mouth caused him endless agony. He suffered from cancer of the palate, most of which had to be removed in a series of 33 operations, so he could be forgiven for his attitude towards adults whom he called orally arrested, breast-fixated and infantile simply because, unlike him, they were able to enjoy adult oral pleasures.

Finally, it is important to examine the female lips as a major source of facial signals. The changing moods of their owners affect their position in four different ways: open and shut, forward and back, up and down, tense and slack. Combined in different ways these four shifts give us an enormous range of oral expressions. The changes are brought about by a very complex set of muscles which basically operate as follows:

Around the lips is a powerful circular muscle, the *orbicularis oris,* which contracts to close them. It is this muscle that is working hard when the lips are pursed, or adopt some other tight-lipped expression. It is tempting to think of this as a simple sphincter muscle, but that would be to underestimate it. If the whole muscle contracts, the lips are closed, but if its deeper fibres are activated more strongly its contraction presses the closed lips back against the teeth. If its superficial fibres are more active, then the lips close and protrude forward. So the same muscle, operating in different ways, can produce the softly puckered lips of the lover inviting a kiss or the tense, tightened lips of a woman who expects to be hit in the face.

Most of the other oral muscles work against this central circular muscle, struggling to pull the mouth open in one direction or another. To simplify considerably, the *levator* muscles raise the upper lip and help to create expressions of grief and contempt. The *zygomaticus* muscle pulls the mouth up and back in the happy expressions of smiling and laughter. The *triangularis* muscle draws the mouth downwards and backwards in the glum expression of sadness. The *depressor* muscles pull the lower lips down to help form expressions such as disgust and irony. There is also the *levator menti* muscle, which raises the chin and projects the lower lip forward in an expression of defiance, and the *buccinator,* or trumpeter's muscle, which compresses the cheeks against the teeth. This is used not only to blow musical instruments but also to help with the chewing of food. When experiencing acute pain, horror or agonized rage use is made of yet another muscle, the *platysma* of the neck region, which drags the mouth down and sideways as part of the tensing of the neck in anticipation of physical injury.

Complicating matters further are the various vocalizations that accompany the mouth expressions. These add a degree of mouth opening or closing which introduces a new element into the subtleties of facial signalling. Take, for example, the contrasting faces of anger and fear. The key difference is the degree to which the mouth-corners are drawn back. In anger they push forward as if advancing on the enemy; in fear they are retracted as if in retreat from attack. But these opposing movements of the mouth-corners can operate with the mouth open and noisy or with it closed and silent. In silent anger the lips are tensely pressed together, with the mouth-corners forward; in noisy anger – roaring or snarling – the mouth is open exposing both upper and lower front teeth, but again with the mouth-corners in the forward position, making a roughly square-shaped mouth aperture. In silent fear the lips are tensely retracted until they form a wide horizontal slit, with the mouth-corners pulled back as far as they will go; in noisy fear – gasping or screaming – the mouth is opened wide, stretching the lips up and back at the same time. Because fear is retracting the lips, the screamer exposes the teeth far less than the snarler.

Happy faces also have closed and open versions. As the lips pull back and up they may stay in contact with each other, resulting in a wide silent smile. Alternatively they may part to give the broad grin in which the upper teeth are exposed to view. If the sound of laughter is added and the mouth is opened wide the lower teeth may also come into view, but because of the upward curve of the stretched lips these lower teeth are never exposed as fully as the upper ones, no matter how raucous the laughter becomes. If a laughing woman *does* expose her lower teeth fully we may doubt the sincerity of her vocal expression.

Another feature of the happy face is the skin creasing that appear between the lips and the cheeks. These diagonal lines, caused by the raising of the mouth-corners, are the naso-labial folds, and they vary considerably from individual to individual. They help to 'personalize' our smiles and grins, an important visual factor in strengthening bonds of friendship.

There is one contradictory face, the sad smile, that illustrates another subtlety of female expressions, namely the ability to combine seemingly incompatible elements to transmit complex moods. In the sad smile the whole face composes itself into the twinkly-eyed look of good humour – except the corners of the mouth, which stubbornly refuse to jack themselves up into the appropriately raised position. Instead they droop to create the 'brave smile' of the harassed hostess or the sardonic smile of the schoolmistress refusing a request. There are many other 'mixed' or blended expressions that, together with the single-minded ones, provide the female face with the richest repertoire of visual signals in the animal world.

9. THE MOUTH

The female mouth works overtime. Other animals use their mouths a great deal – to bite, lick, suck, taste, chew, swallow, cough, yawn, snarl, scream and grunt – but the human female has added to this list. She also uses it for talking, smiling, laughing, kissing, whistling and smoking. It is hardly surprising that the mouth has been described as 'the battle ground of the face'.

Inside the lips, the mouth contains an essential feature – the tongue. Without their tongues women could not talk and would be robbed of one of their supreme qualities – the ability to communicate verbally better than any other animal in the world and better even than the human male. Brain scan studies have confirmed what many have suspected, namely that women are, by nature, more fluent talkers than men. This is an evolutionary statement, not a cultural one. When posed a verbal task, much more of the female brain is employed in registering an answer than is the case with the male. Primeval women were the organizing communicators of tribal life (while men, out on the periphery, were stealthily hunting down prey with barely a grunt to break the silence) and today's women have inherited this quality to their great advantage.

The tongue's role in the act of speaking is sometimes under-estimated, the larynx being given all the credit, but this error is quickly corrected by trying to speak with the tongue held on the floor of the mouth. Anyone who has visited a dentist will have discovered this.

The tongue is also, of course, a major player in the act of feeding, being actively involved in tasting, masticating and swallowing. Its rough upper surface is covered in papillae that carry a total of between nine and ten thousand taste buds. These are capable of detecting four tastes: sweet and salt on the tip of the tongue; sour on the sides of the tongue; and bitter at the back of the tongue. It used to be thought that *all* tasting took place on the upper surface of the tongue, but it is now known that this is not the case. There are sweet and salt taste buds elsewhere in the mouth, especially on the upper throat, while the primary tasters of sour and bitter are on the roof of the mouth at the point where the hard palate meets the soft palate.

It is believed that these particular taste responses exist because it was important for our ancestors to be able to tell the ripeness – and therefore sweetness – of fruits; to be able to maintain a correct salt balance; and to be able to avoid certain dangerous foods – which would have tasted very strongly bitter or sour (acidic). All the subtle tastiness of our foods derives from a mixture of these four basic qualities, aided by other flavours which we smell in the nose.

Besides taste, the surface of the tongue is also responsive to food texture, to heat and to pain. During mastication it rolls the food round and round in the mouth, testing it for lumps. When it judges that all sharp pieces have been crushed or rejected it participates in the crucial act of swallowing. To do this its tip presses against the top of the mouth and then its rear part humps up to catapult the wad of saliva-soaked food into the throat and on its way to the stomach. This extremely complex muscular action is something that is taken entirely for granted because it is so automatic. It is so basic, in fact, that babies are able to perform the action well in advance of any need for it – while they are still inside the womb.

When the meal is over, the tongue busies itself like an outsize toothpick, flicking this way and that, trying to dislodge annoying particles of clinging food from between the teeth.

Because of its protected position inside the mouth, the tongue has rarely been the subject of any cosmetic 'improvements'. However, towards the end of the twentieth century it finally met its match, when the female mouth suffered a strange new intrusion in the form of tongue-studs. As part of a youthful search for adult disapproval, young females subjected themselves to the pain of having their tongues pierced for the insertion of metal studs. Despite the fact that this hindered the clarity of their diction, this form of mutilation was even endorsed by certain pop singers.

Apart from its role as a symbol of social rebellion, the tongue-stud seemed to offer only one advantage. According to one male partner of a stud-wearer, deep kissing without one was like beef without mustard.

An unforeseen disadvantage is one that was discovered in the summer of 2003 when an English woman on holiday in Corfu was struck by lightning. The lightning bolt was attracted by the metal stud in her tongue, shot through her body and departed through her feet. She nearly died – her tongue was badly blistered, her body shook for 10 minutes, she was temporarily blinded and she was unable to speak for three days. As she said later, she had needed the vacation to recharge her batteries but her tongue-stud had taken this too literally.

Inside the lips are the teeth which, in the human species, are employed almost exclusively for feeding. A woman may occasion-ally use them to snip a thread of cotton, but their non-feeding uses are much rarer than in other species. Give an ape a strange object to examine and it will pick it up and almost immediately raise it to its mouth to explore it with lips, tongue and teeth. It may then manipulate it with deft fingers, but overall there is a dependence on both digital and oral contact, with the oral playing the major role. This is also true of human infants, whose parents must always be on the watch for dangerous objects being thrust into tender mouths.

As we mature, however, the mouth gradually loses its 'investiga-tive role', which is taken over almost exclusively by our superior hands. This switch also applies to fighting. Apes in a rage grab at their opponents and bite them. Humans in a rage beat their oppo-nents over the head, and punch, kick and wrestle with them. They bite only as a last resort. The same is true of killing prey. Again the hands – with the help of weapons – have taken over the task of the lethal bite so common among carnivores. Along with this shift from mouth to hand, the human teeth have become smaller and are rather modest compared with those of other species. Our canines have ceased to be fangs with long sharp points. They are only slightly longer than the other teeth, with no more than small blunt points to remind us of our distant ancestry.

The full complement of adult human teeth is 32, 28 of which establish themselves by puberty, having gradually replaced the smaller set of 'milk teeth' which we use during childhood. The last four teeth, the wisdom teeth at the back of the mouth, emerge as we become young adults. Sometimes a few of these, or all of them, fail to appear, so that adult mouths can vary in numbers of teeth from 28 to 32.

There are slight differences between male and female teeth, especially in the front upper teeth. A woman's teeth usually curve more gently than a man's. His are generally more angular and blunt. Also, because women have lighter jaws than men, their teeth tend to be slightly smaller on average.

Apart from their obvious actions of biting and chewing food, teeth are also said to clench and clamp, to gnash, grit and grind, and to chatter with cold. Clenching or clamping the teeth occurs at moments of intense physical effort or when someone is anticipating pain. It is seen on the face of the grappling wrestler or the child about to be injected and is a primeval response to possible injury. If a blow were to fall on the face of an open-jawed individual it could cause much more damage, clashing the teeth together and possibly splintering them or dislocating the 'unclamped' lower jawbone.

To gnash or grit the teeth is the same as to grind them, and it is hard to understand why the language needs three words for the same action, especially as it is so rarely used in real life. During the sleep of many individuals, however, a gentle grinding together of the teeth does apparently take place, indicating a kind of suppressed anger. Again, this is a primitive response which re-surfaces as a sort of 'muscular dreaming', with the frustrated individuals symbolically grinding down their enemies in the safety of their slumbers.

Although tooth enamel is the hardest substance in the entire human body, dental decay is the most common human ailment in the world today. The cause seems obvious enough. A bacterium in the mouth, *Lactobacillus acidophilus* loves carbohydrates, and if particles of sugary or starchy foodstuff are left clinging to the teeth or gums after eating it rapidly ferments them into lactic acid. The bacterium loves this acid even more and starts to reproduce wildly, dramatically increasing the whole process until the saliva in the mouth has become unusually acidic. The acidity then eats away at the surface of the teeth, making small holes in the enamel that develop into festering cavities. All this has been confirmed in a number of ways. For example, children growing up in wartime Europe, when there was very little refined sugar or starch, had fewer cavities. Also, animals fed with a sugar-rich diet develop tooth decay if they eat their food in the usual way but do not if the same diet is tube-fed to them and never touches their teeth. Furthermore, wild chimpanzees living deep in the forest have excellent teeth, while those scavenging near human settlements have rotten teeth.

Yet there are some strange facts about tooth strength that we simply do not understand. Some individuals, for example, seem to be almost immune to decay even when they eat the sickliest of sweet diets. Others fall prey to decay despite great care with both diet and tooth-cleaning. Also, logic would insist that the lower front teeth would, with the help of gravity, become the most food-laden and therefore acid-attacked. Surprisingly, these are the most decay resistant of all the teeth. In the Western world nearly 90 per cent of people have healthy decay-free lower front teeth. In sharp contrast more than 60 per cent have lost their upper-right-middle molar from decay. Despite great advances in dental science the teeth still retain some of their mysteries.

Western eyes have always regarded a healthy set of gleaming white teeth as an essential mark of beauty, but many cultures have taken a different view. One trend has been to remove the central incisors in order to emphasize the pointed canines, which makes the mouth look more menacing and beast-like – almost a Dracula face. This technique has been employed in parts of Africa, Asia and North America.

Another method of making teeth look savage is to file them to sharp points. This too has occurred over a wide range, from Africa to Southeast Asia and the Americas. Sometimes precious stones or metals were inlayed in the teeth to add glamour to them as high-status displays. Many of these tooth operations and mutilations were carried out at special times in the lives of the tribespeople, especially at puberty and at marriage, implying that the mouths were being used symbolically as 'displaced genitals'.

In some areas the impact of teeth was reduced rather than exaggerated. In Bali, for instance, young adults were subjected to painful tooth-filing to flatten out the points of the canines and make the human mouth look *less* like an animal mouth. In certain other Eastern cultures the females blackened their teeth or dyed them dark red, making them virtually disappear from sight and creating an infantile expression, as though they had suddenly regressed to the gummy stage of babyhood. In this way they made themselves appear more subordinate and submissive to their males.

Because, in the Western world, flashing a wide expanse of the whitest of white teeth is today considered an essential part of feminine beauty (a beauty that can now be enhanced by modern bleaching techniques), it is hard for Westerners to comprehend the idea of black teeth as appealing. White is, after all, the natural colour of healthy young teeth, so how could blackening them become linked with an attractive visual display?

The answer, in the times of Elizabeth I, was the price of sugar. Only the very rich could afford to fill their mouths with sugar candies, thereby rotting their teeth and discolouring them. It followed that, if you were too poor to rot your teeth in this way, you had to pretend to have done so. And so it bizarrely came about that blackening your teeth gave you a high-status look that enhanced your beauty in social terms. After all, the Queen herself had black teeth – from eating a surfeit of sugared violets and confits.

Black teeth were also considered fashionable in early Japan. They were dyed this colour as part of the elaborate make-up used by high-ranking women. The black teeth (called *ohaguro*) were said to make a lady look especially beautiful. The dye was prepared by soaking iron filings in sake or tea. Its use reached a peak in the eighteenth century and continued into the nineteenth until, in 1873, the Empress was seen to be displaying white teeth, at which point the fashion for black teeth went into a rapid decline.

In other parts of the Orient, the chewing of betel was a popular indulgence that also led to blackening of the teeth. Betel leaves, palm nuts and lime paste made from ground seashells were mixed together to form a wad that was used rather like chewing tobacco. Bits of nut were covered in the paste and then this was wrapped in the leaves of the betel vine. This package was called a quid. Pushed into the side of the mouth and chewed repeatedly, it acted as a mild stimulant that also reddened the lips and blackened the teeth. Its use was so widespread in Southeast Asia that local girls would exclaim, 'Only dogs, ghosts and Europeans have white teeth.' Its popularity began to decline in the twentieth century, first in the cities and then later in the country areas.

Betel chewing usually only created dark brown teeth, and in some countries – Vietnam for instance – ladies who wanted to perfect their beauty by having teeth that were jet black had to take special steps to ensure this. Painting the teeth with black lacquer was the answer, but it was not a simple matter because the mouth's saliva kept washing the lacquer away. Because of this the application of the lacquer had to be turned into a special ceremony involving several treatments and special restrictions including no solid food for a week and liquids taken only through a straw. For teenage girls it was a coming-of-age ritual, after which they were considered beautiful enough to be married. If asked what was wrong with white teeth, they would reply that such a condition was fit only for savages and wild animals.

Towards the end of the twentieth century, modern women in the Western world showed the first signs of interfering with the pure white surfaces of their teeth. There were no black teeth on view, however, the new fashion trend being for 'tooth jewels'. Pioneers of this fad went to the lengths of having small holes drilled in their teeth to hold tiny diamonds. The gleaming smile became a dazzling one. But this procedure was too drastic for most women and the fashion remained an extreme one. Then certain celebrities, including one of the high-profile Spice Girls, took the radical step of displaying one gold tooth. Soon it was possible to have a non-permanent gold tooth-cap fitted. Then the fashion for gluing very small jewels on to fingernails spread to the mouth and temporary tooth jewels suddenly became popular. As one breathless advertiser put it:

'Sometimes known as razzle-dazzle, ghetto glow, tin grin or metal mouth, tooth jewels have recently become all the rage.' Their appeal is that their attachment, using dental glue, takes only three minutes, and they can easily be removed later on if their owners become bored with them. Tiny crystals in the shape of hearts, flowers, circles or stars, between 2 and 4 millimetres in size, are used and are kept in place from as little as one day to as much as a year. They vary from flashy to discreet, according to which teeth they are attached to. Although decorative, the fact that they break up the broad white expanse of tooth-smile will probably mean that they are no more than a passing fad.

The two main features of the mouth – the teeth and the tongue – are kept moist by the secretions of three pairs of salivary glands. The pair embedded in the cheeks are known as the parotid glands and they produce about a quarter of the saliva; those beneath the jaw under the molar teeth – the sub-mandibular glands – are the most productive, accounting for about 70 per cent; and those beneath the tongue – the sub-lingual glands – contribute the other 5 per cent. Estimates of a person's total output of saliva per day vary (between 1 and 3 pints). More food means more saliva. Fear and intense excitement mean less saliva.

When saliva leaves the ducts of the salivary glands it is free of all bacteria, but by the time it has swirled around the mouth a few times it will have collected up between 10 million and 1,000 million bacteria per cubic centimetre. It acquires these from the tiny fragments of 'wet dandruff' that are always present inside our mouths, as the skin surfaces there repeatedly slough off old layers and replace them with new tissue.

Saliva has a number of functions. It moistens the food as it enters the mouth and makes it accessible to the taste buds, for dry food cannot be tasted at all. It also lubricates the wad of chewed food before it is swallowed and in this way eases its passage down the oesophagus. Its quality as a lubricant is improved by the presence of a protein called mucin. If food is chewed for any length of time an enzyme in the saliva called ptyalin starts breaking down starch to maltose. Ptyalin also acts as an oral germ-killer, as do other lyzozymes that help to clean the mouth and teeth. Saliva also contains chemicals that create slightly alkaline conditions that help to reduce acid attack on the tooth enamel. Finally, the lubricating action of saliva improves the quality of vocal tones, as anyone who has tried croaking with a dry mouth will appreciate.

10. THE NECK

In the West, men tend to look upon the female neck simply as something that holds up a woman's head. They may be aware that the skin of the neck is sensitive to gentle caresses and that kissing it softly can arouse the female partner during sexual foreplay (giving us the term 'necking'), but beyond that it is not credited with much importance. It is certainly not viewed as a major erotic zone.

The situation is very different in Japan, where exposing the back of the female neck is considered to be one of the most sexually tantalizing actions possible – the equivalent of exposing the breasts in the West. It is an action expected of a Geisha, but shunned by a respectable Japanese wife, who would rather fit her collar snug to the nape of her neck.

Traditionally, every Geisha was trained in the art of elegantly exposing her neck and this can still be witnessed on the bodies of the few traditional Geishas remaining in Kyoto today. Their costumes are high at the front and low at the back, with the collars pulled down to expose the skin of both the neck and the upper back, 'well below the first large bone of the spine'. As one commentator remarked, men everywhere seem to enjoy a plunging neckline, but in Japan it plunges backwards.

When applying her smooth, white make-up (which includes the vital ingredient of nightingale droppings), a Geisha will leave a small margin of bare skin around her hairline. This emphasizes the artificiality of the make-up and excites a man by drawing attention to her bare skin beneath the mask of white. According to one observer, the erotic significance of this custom is enhanced by the special shape of the 'naked neck' region, the nape showing 'a perfect V of bare skin, which hints at a woman's private parts'.

There was a special Japanese phrase to describe the beautiful shape of the hairline on the back of the neck – *komata no kereagatta hito* – but its meaning has changed. Because the make-up is deliberately applied to mirror the shape of the genitals, the phrase now signifies 'a Geisha with a lovely genital area'.

An intriguing suggestion has been made to explain the Japanese switch from breasts to neck, as a focus of erotic attention. It is pointed out that, traditionally, Japanese children spend more time strapped to their mother's back than they do nursing at the breast. This, and the fact that the breasts of Japanese women are comparatively modest in their dimensions, is thought to be the reason for the neck-fixation.

Anatomically, the neck has been described as the most subtle part of the human body. Besides containing the vital connections between mouth and stomach, nose and lungs, and brain and spine, it houses the crucial blood vessels between heart and brain. And surrounding these connection lines are complex groups of muscles that enable the human head to dip and nod, shake and twist, turn and toss and perform a whole range of movements that convey important messages during social interactions.

Traditionally, the exceptionally feminine figure is endowed with a graceful 'swan-like' neck while the exceptionally masculine figure is 'bull-necked'. These differences are real enough. The female neck is longer, more slender and more tapered and the male neck is shorter and more thickset. This is partly because the female has a shorter thorax, the top of her breastbone being lower in relation to the backbone than the male's, and partly because of the stronger musculature of the male. This gender difference undoubtedly developed during the long hunting phase of human evolution, when males with stronger, less-snappable necks were at an advantage in moments of physical violence.

Another gender difference in the neck concerns the Adam's apple, which is much more conspicuous than Eve's equivalent. This is because women, with higher pitched voices, have shorter vocal cords requiring a smaller voice box. Female vocal cords are only about 13 millimetres (0.5 inches) long, while those of the male are 18 mm (0.7 inches). The female larynx is roughly 30 per cent smaller than that of the male and is also placed slightly higher in the throat, which has the effect of making it less prominent. This laryngeal gender difference does not appear until puberty, when the male voice deepens, or 'breaks'. The adult female voice and the female larynx are essentially more infantile than the male's, retaining a pitch of between 230 and 255 cycles a second, while the adult male voice sinks to somewhere between 130 and 145 cycles a second.

For some reason experienced prostitutes have been found to have a larger larynx and deeper register than other women. Why their occupation should make them vocally more masculine is not clear, although it suggests that their unusual sexual lifestyle may in some way disturb their hormonal balance.

Because the female neck is more slender than the male's, artists have frequently exaggerated this feature to create super-feminine images. Cartoonists portraying attractive women invariably narrow and lengthen the neck even further than normal anatomy would permit. Also, model agencies, when selecting individual girls for training, choose those with necks that are thinner and longer than the average.

In one culture this desire for long-necked women has been taken to remarkable extremes. The Padaung branch of the Karen people of upland Burma boast what came to be known in Europe as 'giraffe-necked' women. The word 'padaung' means 'brass wearer' and the females of this group are required by local fashion to start wearing brass neck-rings from an early age. To begin with, five rings are fixed around the neck and this number is increased gradually, year by year. As adults, the total number displayed is usually between 20 and 30, but the ultimate goal is to reach 32 – a feat seldom realized. Brass rings are also put on the arms and legs, so an adult female might carry around with her a weight of brass totalling 20 and 30 kilos (between 50 and 60 lbs). Despite this encumbrance the women of the tribe are expected to walk long distances and work in the fields.

The most amazing aspect of this custom is the extent to which it artificially lengthens the women's necks. The record neck-length documented is 40 cm (15.75 inches). The neck muscles are stretched so severely by this practice that the neck vertebrae are pulled apart in a completely abnormal way. It is said of such a woman that if her heavy brass rings were removed, her neck would be unable to support her head. Europeans, fascinated by this dramatic cultural distortion of the human body, paraded a number of these long-necked women in circus sideshows – until human displays of this kind were no longer considered socially acceptable.

For the Padaung women themselves the chief concern today is not, as one might imagine, the physical distortion of their bodies or the restriction of movement imposed on them by their bizarre adornment but, more mundanely, how they can find enough money to pay for the expensive brass rings. A recent solution has been to slip over the border into Thailand where they can charge $10 a time to have their photographs taken with tourists. To some observers this seems like a deplorable return to the circus sideshows of old, but it can equally be argued that, given the rising cost of brass rings, it does at least keep an ancient tribal custom alive.

If you ask tribal historians how the Padaung long-neck custom first began, they will tell you that, in ancient times, women were threatened by attacks from marauding tigers and that they started wearing thick neck-rings to protect themselves. Modern Padaung women ignore this legend, saying simply that they adopt these decorative extremes because it makes them look more beautiful. And who are we, in the West, with our tongue-studs, pierced navels and genital rings, to criticize them?

In occult circles the neck has always been a body zone of major importance, and it is no accident that, in vampire mythology, the ritual bite is always located at the side of the neck. In some cults, such as voodoo in Haiti, it is believed that the human soul resides in the nape of the neck, and it was the occult significance of the neck that led to the widespread use of necklaces in earlier days. They were more than mere decorations, having the special function of protecting this vital part of the human anatomy from hostile influences such as the 'Evil Eye'.

The oldest known necklace is one that was worn, not by a modern human, but by a Neanderthal. Indeed, necklaces are a truly ancient form of body decoration. Of two prehistoric ones, both found in France, one from La Quina, made of animal teeth and bone beads, has been dated to 38000 BC, and one from the Grotte du Renne, made of grooved and notched animal teeth beads, has been recorded as 31000 BC. And in Western Australia, at a site called Mandu Mandu, another amazingly early necklace has been dated to 30000 BC. Finally, at Patnia, in the Maharashtra region of India, a 23000 BC necklace of disc beads, manufactured from Olivia shell and ostrich eggshell, has been discovered. These few examples clearly show that the wearing of a necklace of some kind was not an isolated local custom, but one that was already remarkably widespread 30 millennia ago.

Some of the very first necklaces were made from simple objects such as fish vertebrae, but an exceptional one found in France, that was created over 11,000 years ago, in the Old Stone Age, was made of 19 beautifully carved bone-fragments. Eighteen of them were elegantly fashioned as ibex heads and one as a bison head. It demonstrates very clearly how much attention was given to artefacts worn in the neck region.

The neck also became the focus of certain occult ritual practices. It was discovered that by applying pressure to the large carotid arteries which run up the side of the neck, carrying blood to the brain, a subject could be made dizzy and confused – an easy prey to suggestion. What was happening, of course, was that the subject's brain was being deprived of oxygen; but in the mumbo jumbo of religious rites, her condition could conveniently be attributed to the supernatural.

A much healthier form of neck manipulation was developed by Matthias Alexander, who founded what was to become known as the Alexander Technique. This was based on the idea that by modifying the basic posture of the neck on the shoulders, it was possible to cure not only certain physical symptoms but also a variety of psychological disturbances. Some critics have argued that this concept gives the neck an almost mystical power over the rest of the body, but there is a simpler explanation. Because urbanites spend so much time hunched over a desk or a table or slouched in a chair the neck gradually loses its natural vertical posture. If through Alexander training this posture can be re-established, the rest of the body automatically follows suit and recovers its correct balance. The scene is then set for a return to a healthy body tonus, which may in turn lead to a healthier mental state. It is really no more mystical than the kind of posture training a ballet dancer receives. In both cases the neck seems to be the key that unlocks the body's poise.

Turning to gestures, there are comparatively few that focus specifically on the neck. The most widespread is the throat-cut mime, in which the gesturer uses her hand as a mock knife to slice across the front of her throat. This has two closely related meanings. If done in anger it can indicate what the gesturer would like to do to someone else. If done as an apology it shows what the gesturer feels like doing to herself. In a different context, done by an actress when a scene is going wrong, it means simply 'cut!'

Also widespread is the 'mock self-strangling' gesture in which the woman's hand or hands clasp her own neck and pretend to throttle it. As with the mock throat-cut this has two closely related meanings, signfying either 'I want to strangle you' or 'I could strangle myself'.

Another popular neck gesture is the 'I-am-fed-up-to-here' signal, in which the forefinger-edge of a palm-down hand is tapped several times against the throat. The implication is that the gesturer has been stuffed so full of something that she cannot take any more.

More important than these regional gestures are the many neck actions that result in head movements or postures. These are of two kinds. First, there are actions that adjust the woman's body to her environment, as when her head is turned to look at something, cocked to listen to a sound, or raised to sniff the air. Second, there are the actions whose sole function is to transmit visual signals to companions. These include such movements as the head nod, bow, shake, dip, toss and point. In these and most other neck actions, there is no difference between males and females, but there are three cases where a specifically feminine signal is transmitted.

One is the *head beckon,* in which the woman's neck pulls her head away from her companion, jerking it backwards with a slightly tilted posture. This gesture says 'come with me' or 'come over here' and is a substitute for the hand or forefinger beckon. It is most likely to occur when the beckoner wishes to signal without being too conspicuous. It is the traditional head movement given, almost imperceptibly, by a streetwalking prostitute to a potential client, when he is hesitating to approach her. Today it is sometimes also used between established couples as a joke invitation to sex, with the female partner being provocative by 'acting the whore'.

Another is the *head lower* in which the woman's neck tips her head down and keeps it in its drooped posture. This is a way of cutting off the outside world, but because it embodies a height reduction it has a subordinate air about it. The head swept away to one side can be haughty but the lowered head cannot. Modesty and shyness are signalled by this sudden lowering of the head to hide the face, and feminine coyness can be indicated by a combination of the head lower and an upward glance of the eyes.

A third neck movement, often seen when a woman is in a friendly or loving mood, is the *head cock*. In this, she tilts her heads to one side and holds it there. It is done while she is facing her companion, at a short distance. This action derives from a childhood comfort-contact in which, as a little girl, she once rested her head against the body of her protective parent. When, as an adult female, she tilts her head to one side, it is as if she is leaning it against a now-imaginary protector. This 'little child' act is contradicted by the mature sexual signals of the adult body of the gesturer, giving the head cock an element of coyness. If used as part of a flirtation, the head cock has about it an air of pseudo-innocence or coquettishness. The message says, 'I am just a child in your hands and would like to rest my head on your shoulder like this.' If used as part of a submissive display the gesture says in effect, 'I am like a child in your presence, dependent on you now as I was when I laid my head on my parent's body.' It is only a mild signal, however, and does not make this point strongly but merely hints at it.

There are many more head movements and postures created by the human neck muscles as specific social signals, but the few mentioned here are already sufficient to illustrate the subtlety and complexity of the ever-shifting neck. Anyone who has suffered from a stiff neck or been forced to wear a medical collar following neck injury will know just how deprived the human frame feels when it cannot express itself with this part of the body.

11. THE SHOULDERS

Female shoulders are rounder, softer, smoother, narrower and thinner than those of the male. They may not be as powerful as the broader masculine shoulders, but their smoothly rounded shape – resulting from a subcutaneous pillow of fat – does give them an erotic quality whenever they appear unclothed. And an off-the-shoulder style of dressing has the added appeal that the clothing promises, at any moment, to slip down to reveal the breasts.

Because the smoothly curved 'corners' of the exposed female shoulders display almost hemispherical patches of flesh, poetically described by one author as 'two rounded orbs, one erotic pearl on each side', they cannot avoid transmitting the primeval female sexual signal that originates in the hemispherical shape of the buttocks. This 'paired hemisphere' signal, which has such a powerful impact on sexually responsive males, finds body-echoes, not only in the female breasts, but also in the female knees and shoulders, when certain postures are adopted. When a young woman with tightly bent legs hugs her knees close to her chest, the knees, if exposed, present a pair of smooth hemispheres to watching male eyes. In the same way, if unclothed shoulders are hunched up they, too, can echo the 'paired hemisphere' signal, giving them added appeal to male eyes. In addition, a typical 'glamour' pose involving the resting of the chin on one raised, naked shoulder, emphasizes and draws attention to the roundedness of the smooth flesh. In these ways, shoulders, even though they lack any primary sexual function, can transmit mild erotic signals.

Before examining the ways in which various cultures have modified the natural female shoulder-line, it is worth taking a brief look at the biology of this part of the female anatomy.

The main function of the shoulders is to provide a strong foundation for the multi-purpose arms. Ever since our ancestors adopted an upright way of life, our 'front legs' have become increasingly versatile, and the shoulder girdle, or pectoral girdle, has had to serve that versatility by becoming more flexible. The collar bones and shoulder blades are capable of movements through about 40 degrees and, with their complex muscles, can help the arms to swing, twist, lift and rotate in an amazing number of ways.

The shoulders of an average female are seven-eighths as broad as those of an average male. Even more important is their measurement from front to back. In this direction the difference is greater, reflecting the comparative weakness of the female shoulder musculature.

Inevitably this sex difference led to a variety of cultural exploitations. If feminine shoulders are narrow, then making them narrower still should increase the femininity of the woman in question. However, although this type of exaggeration is possible in other parts of the female anatomy, it is difficult in the shoulder region, and has rarely been attempted. One exception is illustrated in John Bulwer's seventeenth-century volume of anthropology, called *A View of the People of the Whole World,* where he shows a young woman with her shoulder blades squashed together to an abnormal degree. He writes that 'Narrow and contracted shoulders were esteemed so proper to women of old, that they affected this composure of the shoulders, and learnt it very diligently as a great elegancy and beauty . . . A handsome slender woman [with], as it were, pinioned shoulders.'

In complete contrast, women who have wished to assert themselves have adopted artificially broadened shoulders, and this has happened at several points in the recent past. It was noticeable in the dress of the Emancipated Woman of the 1890s. In her bid for sexual equality she displayed her mood by adopting 'shoulder equality'. Fashion historians have recorded the shift: 'The slightly puffed shoulders developed into epaulettes and then into something looking like small bags until by 1895 they were rather like a pair of large balloons quivering on the shoulders.' These broad-shouldered women competed with men by taking university degrees, going out to work and engaging in sports that had hitherto been barred to them. Beneath their masculine attire, however, they still wore corsets and petticoats. They were masculine in public but feminine in private.

The second wave of big-shouldered women appeared in the 1940s, during the Second World War, when rather square-cut, military-style clothing was adopted even by civilians. This included a stiffly supported shoulder-line which often extended well beyond the natural end of the shoulders. It was an appropriate display for a wartime period when women were playing a bigger role in hostilities than ever before.

The third wave arrived in the 1970s with the women's liberation movement. Initially it took the form of what could be described as 'terrorist chic'. Pseudo-battledress tops with epaulettes created the required air of female toughness, and once again the shoulders were squared to give an air of masculine strength. An accompanying shift in glamour figures could also be detected. Leading ladies in films no longer minced and wiggled, they strode out. Girls with naturally wide shoulders were given chances that would have been denied them in the 1960s or earlier. Figures with broad shoulders preferred to be thought of as one of the 'guys'. As an extension of this trend, female body-building surfaced and found a considerable following. A few decades earlier a female musclewoman would have been viewed as a circus freak, but in the feminist climate she became a symbol of the new female strength and had the powerfully developed shoulders to prove it.

The 1980s saw the arrival of female 'power dressing'. This was the decade in which the earlier feminist battledress gave way to the dark business suit. Writers of the period referred to these suits as having 'Joan Crawford Shoulders', harking back to the severe female shoulder-padding of 1940s. But these shoulder-pads were even more exaggerated as the tough new generation of hard-talking female executives began to storm the boardroom. The 1980s shoulders made such an impact that they had the journalists of the day competing with one another to coin new phrases. 'Shoulderism,' said one, 'is making it difficult to find space in an elevator.' Everyone is wearing 'don't push me shoulders' said another.

Other comments in the mid-1980s included: 'Shoulder-pad factories in the Bronx are laying down new assembly lines after years of idleness'; women are so aggressive now that 'they have returned to a wartime shoulder-shape'; women are demanding '*Dynasty* frocks with aircraft-carrier pads'; models with naturally wide shoulders 'are being favoured today', and the singer Grace Jones had 'a haircut that even gave her shoulders on her head'; big-shouldered women are 'toughies demanding their own space'; women now prefer to wear 'those vast cantilevered Star Trek jobs which soar inches up and away from the body'; and finally 'women can never go home again – they'd never get their shoulders through the door'.

As the 1980s gave way to the 90s, the female shoulder softened again. The feminist movement (at least in the West) had made sufficient strides for women to enjoy equality as females, rather than as pseudo-males. The shape of the shoulder now depended on the design of a particular costume, rather than on an all-embracing social statement.

Interestingly, although by the 1990s women had a greater freedom to dress as they liked, the concept of broad-shouldered women lived on as a verbal label, even though it was no longer a costume reality. As late as 1994 an article was written on the subject of the increasing domination of the publishing industry by female executives, under the title 'Why shoulder-pads are back in power'. By this date, real shoulder-pads would have looked rather old-fashioned, but the term itself managed to survive as a metaphor for the triumph of women in a male world.

One aspect of male shoulders that has been difficult for women to imitate is their height above the ground. The average male is about 13 cm (5 inches) taller than the average female, with the result that males have always been able to offer a shoulder to cry on, not because it is broad but because it is high enough to act as a comfortable resting place for a distraught female cheek. With tears and vulnerability out of fashion, the modern female still finds herself faced with a tall shoulder. Since the extra height of her male companions evolved through their primeval hunting activities it seems unfair to many women that modern-day deskbound, pen-pushing males should still display this badge of physical superiority. Unfortunately evolution works at a very slow pace. Another million years of pen-pushing may correct matters, but in the meantime male shoulders will remain stubbornly at female head-resting height. Short of cutting down men's legs, the only hope of shoulder-height equality is the wearing of five-inch lifts on female shoes. The difficulty with this is that very tall shoes make for instability and the need for a helping male hand, which defeats the object of the exercise. For the time being it looks as if, physically, women will have to continue to look up to men, even though mentally they have adopted a very different viewpoint.

The mobility of the human shoulders is such that even when they are not involved in arm movements they can be raised and lowered, rounded and squared, hunched and shrugged. Some of these movements have become modified as special signals in the language of the body, but to understand them it is necessary to look at the more primitive reasons for adopting one type of shoulder posture rather than another.

Basically, shoulders are kept *down and back* in a mood of calm and alertness and are brought *up and forward* in moments of anxiety, alarm or hostility. The cheerfully resolute, dominant woman keeps her shoulders lowered and squared. Women who are being dominated or who are frightened or angry tend to hunch up their shoulders as an act of self-defence. If someone is threatening to hit a woman over the head she automatically tries to protect her head and neck region by tucking her head down into her shoulders, and this tensed up position has become synonymous with unpleasantness of every kind.

It follows from this that if a woman has a stressful day, full of disappointments or irritations, she will keep on tensing up her shoulders. These actions might be useful if she was being beaten with a stick, but they are useless when she is being 'beaten' instead with harsh words. At the end of such a day she will find herself slightly more round-shouldered than she was in the morning when the day's events began. If this is repeated day after day, week after week, she can eventually develop a marked stoop in her posture, with hunched, permanently rounded shoulders. The long sloping neck she displayed as a child slowly shrinks down into her shoulders until it all but disappears. By old age, her chin has come to rest on her chest.

Successful women (that is to say, successful to themselves as well as to the outside world) do not undergo this gradual hunching-up decline, and there have been plenty of ram-rod-straight ninety-year-olds to prove the point. Full of self-confidence and optimism, there are too few blows in their lives to make them duck in anticipation into a life-long hunch. For others, and they are in the majority, there are too many anxieties in modern living to avoid at least some degree of shoulder tensing during waking hours.

Two special shoulder signals owe their origin to this defensive hunching. They are the 'Shoulder Shake' and the 'Shoulder Shrug'. The Shoulder Shake is a conspicuous addition to social laughing. If something makes us laugh when we are on our own we let out a guffaw or a snigger but we do not usually add body movements. These are kept for social occasions, when we are not only amused but also want to convey our amusement to our companions. Because there a slight hunching up of the shoulders when we laugh we can augment our display of good humour by exaggerating this hunching action, repeating it and enlarging it, so that the shoulders rise and fall rapidly with our laughter.

The reason why people 'shake with laughter' in this way is that the basis of humour is fear. Humour must shock us in a safe way and we signal our surprise and our relief simultaneously by laughing. The hunching of the shoulders that accompanies this vocalization is part of the primeval fear element. The repeated rise and fall of the shoulders when they shake with laughter is saying, in effect, that there is fear present, but it is not serious. If it *were* serious, the shoulders would stay up.

The Shoulder Shrug has a similar origin. In this action, the shoulders are raised into a fully hunched position for a brief moment and then dropped again. The hands are turned palm-up in a manner similar to that seen in begging or imploring, and the mouth-corners are lowered. Sometimes the eyes are deflected upwards as if avoiding your gaze. This combination of actions indicates a momentary lowering of status, a symbolic impotence, a fleeting acceptance of an inability to cope.

Most shrugs are signals of ignorance ('I don't know'), indifference ('I couldn't care less'), helplessness ('I can't help it') or resignation ('There's nothing to be done'). They are all negatives, admissions of inability, and with that inability goes the brief loss of status. As the status momentarily sinks, the shoulders momentarily rise. This formal adoption of a tensed-up posture does not mean that we are seriously stressed or that we necessarily feel inferior to, or threatened by, the speaker who has provoked our shrug. It simply means that we cannot deal with their specific question or comment.

The use of shrugging varies considerably from culture to culture but it always has the same basis. In some Mediterranean countries its threshold of use is very low. One only has to make a passing mention of government restrictions, taxation, or the increase in traffic jams, to produce an immediate, prolonged and silent shrug. This expresses the shrugger's complete helplessness in the face of inconceivable folly. Her shrugging posture says: 'These blows keep falling on my poor shoulders and I raise them like this to protect myself but what good does it do?' In countries farther north, shrugging, like other gesture-replies, is considered impolite and it occurs less frequently, but when it does appear it has similar roots.

Not all shoulder-raising is defensive hunching. There are several kinds that do not involve a protective element. Raising and rounding the shoulders forward, with the arms clasping the front of the body is a form of 'vacuum embrace'. It is the action of hugging ourselves in the absence of someone else to hug. Here the raised and rounded shoulders are mimicking the posture they would adopt if the loved one were present to be truly hugged. Another version of this is the raising of one shoulder to make contact with the chin or the cheek. The head rests against the shoulder, again as if performing a tender action towards a loved one. The shoulder is 'standing in', as it were, for the missing shoulder of the absent loved one.

12. THE ARMS

The arms are the least erotic part of the female body. If a man wishes to touch a woman's body in a non-sexual way – to attract her attention, say, or to guide her in a particular direction – the safest place to make contact with her is her arm. Anywhere else would be too intimate.

It is worth remembering that, in evolutionary terms, the human arms are our front legs. Indeed, to any four-footed creature they must look like a pair of useless limbs dangling in the air. But when our ancestors stood up on their hind legs our forelimbs were dramatically relieved of load-bearing and able to specialize as multi-purpose manipulators. Our front feet turned into sophisticated prehensile grippers and our front legs became their wonderfully mobile servants.

The arms serve in two ways: with power and with precision. If the hands must act forcefully – climbing, throwing, beating, punching – the strong arm muscles, such as the biceps and triceps, ripple and bulge into action. If the thumb and fingers are acting with delicate precision the arm then operates as a mobile crane, moving the hand into the ideal position for the more finicky work to be carried out.

The arm is based on three long-bones: the heavy *humerus* of the upper arm and the lighter *radius* and *ulna* of the forearm. These bones are apparent at the shoulder, the elbow and the wrist, but elsewhere they are embedded in muscles. The two bones of the forearm cross over each other when the hand is rotated into a palm-up position which means that the most relaxed arm position is a

117

palm-down one. For those who cannot remember which bone is the ulna and which the radius, the ulna is the slightly more slender one in line with the little finger, while the radius is the stouter one in line with the thumb.

The main arm-muscles and the movements they create are as follows: The *deltoid* is the bulky muscle that rounds off the top of the arm where it meets the shoulder. Its function is to raise the arm up and away from the side of the body. The *biceps* are the bulging muscles on the front of the upper arm. Their function is to bend the arm. The *triceps* are the powerful muscles at the back of the upper arm, with the function of extending the forearm.

Muscle-building techniques make it possible to pump these arm muscles up to an amazing degree and the bulging arms of competitors in female body-building contests create an impression of immense strength. Many males report that they do not find such displays sexually attractive. The main reason appears to be that the amount of effort obviously involved in developing the arms to this extent implies a degree of self-obsession verging on the narcissistic. The top female body-builder seems to be less interested in the body of a male companion than in the body she sees in the mirror.

Another problem with the over-developed female arm is that it looks too masculine. Typical female arms are shorter, weaker and more slender than typical male arms, so that when body-building enlarges them they inevitably lose their normal feminine qualities.

The longer forearms of the male are seen as reflections of his specialized evolutionary role as an aimer and thrower. As a result of this, men are much better javelin throwers than women. The male world record for this event is 96.72 metres (317 feet), the female 72.40 metres (237 feet). This difference is much greater (33 per cent) than the differences for track events (average 10 per cent).

Another gender difference concerns the elbow joint. In females the upper arms are naturally closer to the flanks than is the case for males. The broader shoulders of males mean that their arms hang down away from the body. When they are allowed to dangle in space they have a strongly masculine flavour, but if a male were to force them close in to his sides while spreading his forearms out away from his body he would appear effeminate. This is because there is a greater elbow angle – greater by about 6 degrees – in the female. So the posture of the arms also provides us with significant gender signals that cannot be ascribed to local cultural conditioning.

If the elbow suddenly encounters a hard object, there may be a numbing, stinging sensation accompanied by considerable temporary pain. This is described as 'hitting one's funny bone', a phrase which is a play on the technical term for the bone involved, the *humerus* of the upper arm. It is the knob at the lower end of this bone that is the 'funny bone', because it is there that the ulnar nerve is exposed just under the skin. It is striking this nerve that causes the stinging sensation and for a moment incapacitates the arm in question.

Another anatomical detail of the arm that deserves brief mention is the much-maligned, much-shaved and much-sprayed armpit. Technically known as the *axilla*, this small hairy zone plays an important role in chemical signalling and reflects a major change in the sexual habits of the human species. When our remote ancestors mated, with the female on all fours, the armpits were nowhere near the partner's face. When eventually we adopted a vertical gait and switched to a face-to-face posture as the dominant sexual position, embracing couples found their noses close to the shoulder regions of their partners. Nearby was the partly enclosed armpit, the ideal site for the development of specialized scent glands. Their abundant presence is unique to the human species and they are present in both sexes.

Females possess more of these scent glands than males and the odours produced differ between the sexes, suggesting that they operate as sexual signals between amorous partners. Indeed, recent experiments have revealed that blindfolded men become more sexually aroused by sniffing female armpit sweat than by smelling expensive commercial perfumes.

These armpit scent glands are called apocrine glands and their secretions are slightly oilier than ordinary sweat. They do not develop until puberty, when the arrival of sex hormones activates them and at the same time causes the growth of armpit hair. The hair acts as a scent trap, keeping the glandular secretions within the axillary region and helping to intensify their signal.

119

There is an old English folk custom, handed down from generation to generation, based on the idea that if a young man wishes to seduce a young woman at a dance he must place a clean handkerchief in his armpit, beneath his shirt, before starting to dance. Afterwards he takes it out and fans her with it as if trying to cool her. In fact what he is doing is wafting his apocrine scent over her in the hope that she will be seduced by its fragrance.

In rural Austria the trick worked the other way. Young women would place a slice of apple in their armpits when dancing and when the music stopped, they would offer it to a favoured male partner to eat. When he obediently consumed it he was automatically exposed to her personal sexual fragrance. This trick was also known in Elizabethan England, where a whole, peeled apple (known as a 'love apple') was placed in the young woman's armpit until it was soaked with her sweat, when she would give it to her sweetheart, who would inhale its fragrance.

Later, in the sixteenth century, the sexual impact of female armpit fragrance is said to have made itself felt at the French Court. A beautiful young princess, Mary of Cleves, the wife of the ugly Prince of Condé, was feeling hot from vigorous dancing at court, and retired to change her sweat-stained chemise in one of the side-rooms, adjoining the Louvre ballroom. The Duc d'Anjou (who would soon become King Henri III of France), also suffering from the heat, entered this side-room and, thinking the discarded chemise was a napkin, used it to wipe his perspiring face. According to the chronicler of the day, his senses were deeply affected by this action. Already a secret admirer of the teenage princess, from the moment he inhaled her fragrance he developed an uncontrollable passion for her and was forced to break his silence and tell her of his intense feelings towards her. It was a doomed infatuation that was to cause him agony and heartbreak in the years that followed.

Considering the major industry that today is based on the sale of underarm deodorants, these stories sound rather odd. If human beings carry such a powerful sexual stimulus under their arms, why do so many go to such trouble to remove it, with washings, rubbings, sprayings and, in the case of women, depilations? The answer has to do with clothing. The young man in the English folk tale, well scrubbed and wearing his best clean shirt for the dance, produces *fresh* apocrine secretions from his scent glands. Soaked in these, his clean handkerchief really does carry a strongly sexual scent signal. That is the primeval system at work. Sadly, today, with our bodies covered in layers of clothes, our sweaty skin can easily become a hothouse where millions of bacteria start to decay. Our natural body fragrance turns sour in this unnaturally confined environment and our scents become stinks. The unpleasantness when this happens concerns us so much that we would rather spray our armpit glands into abject submission than risk our axillary attractions turning into the socially dreaded 'body odour'.

As long ago as the first century BC, the Roman poet Ovid, in his handbook of seduction *The Art of Love*, warned ladies that they 'carry a goat in the armpits'.

Recent research has shown that the armpit secretions of males and females differ chemically in several ways and have odour-appeal specifically directed at the opposite sex. The male's secretion is said to be muskier, with the male hormone androsterone playing a significant role. However, in its pure, fresh form neither male nor female secretion is easily detected consciously by the human nose. They appear to act at an unconscious level, leaving us feeling stimulated but not knowing quite why.

Orientals, incidentally, lack this underarm odour-signalling system almost entirely. Among the Koreans at least half the population have no axillary scent glands at all. The glands are also rare among Japanese – with 90 per cent of the population having no detectable underarm odour. In fact, to have strong-smelling armpits in Japan is considered a disease and has been given the technical name of 'osmidrosis axillae'. At one time individuals suffering from this 'ailment' were even excused military duty. In China the situation is even more extreme, with only 2–3 per cent of the population having any detectable armpit odour. Because of this racial difference, Orientals often find the natural armpit odours of Europeans and Africans overpowering and even offensive.

The removal of armpit hair (by shaving, waxing or creaming) is a comparatively recent practice, first introduced in the Western world in the 1920s by the newly burgeoning cosmetics industry.

Advertisements telling young women that they would be more fragrantly appealing if they eliminated their armpit-hair 'scent-traps' had a powerful impact and before long the majority of Western women had fallen into line. Today it is claimed that fewer than 1 per cent reject underarm depilation as a routine procedure.

Occasionally there has been a minor rebellion against this popular form of armpit 'mutilation'. The famous lovers' guide, *The Joy of Sex*, published in 1972, was strongly opposed, commenting: 'Armpit – classical site for kisses. Should on no account be shaved.' Shaving 'might be forgivable in a hot climate with no plumbing, but now it is simply ignorant vandalism'. It added the curious advice that the armpit 'can be used instead of the palm to silence your partner at climax' – presumably to ensure that your axillary scent signals are appreciated to the full.

It is not clear how many young women abandoned depilation as a result of this advice, although *The Joy of Sex* seemed to think that in the early 1970s there was already a trend in this direction, saying that 'a new generation started to realize that it was sexy' to retain armpit hair. Judging by magazine photographs and films that have appeared in the decades since then, the fashion world has ignored this trend and the majority of young women appear to have followed this lead. When one famous Hollywood actress recently raised her arm to wave to the crowd at a star-studded premiere and revealed that she had a hairy armpit, it was a hot topic for the gossip columns and was generally considered to be repulsive.

Despite this, the closing years of the twentieth century saw the arrival of a magazine called *Hair to Stay,* sub-titled '*The World's only Magazine for Lovers of Natural, Hairy Women*', although it had to admit that it was fighting an uphill battle: 'Women in the 90s who choose not to shave their underarms are categorized, ridiculed and embarrassed. These women are considered to be lesbians, radical feminists, immigrants who don't know any better, or hippies left over from the 60s.' This is wrong, the magazine insisted, because 'from a psycho-social point of view, the act of removing body hair is a rebellion against sexuality'. Underarm hair, it claimed, 'acts as a transmitting antenna, sending out signals to invite sexual intercourse'. Warming to its theme, it went on to say that when an adult woman shows off a hairless armpit she is symbolically offering herself as a child and is therefore encouraging a perverted attitude to sex. It conveniently failed to point out that this extreme line of argument would end up accusing every clean-shaven adult male of encouraging paedophilia – because little boys have no beards.

The simple fact is that removal of body hair by adults, of either sex, makes them look cleaner and younger and helps them to reduce their scent-signalling. Since modern adults, especially in crowded urban conditions, frequently find themselves in unnaturally close proximity with other adults, in totally non-sexual situations, there is every reason to want to damp down the primeval sexual signals. For this reason, it seems likely that body depilations of various sorts will continue to thrive, whatever the social rebels have to say on the subject. Only if we were all to return to a semi-naked tribal condition would their arguments become valid.

Turning to arm postures, there are four main ones to consider: Arms Down, Arms Up, Arms Spread and Arms Forward. The Arms Down posture is the neutral one, with the arm muscles at their most relaxed and inactive. As part of the balancing act of bipedal loco-motion we swing our arms slightly out of this resting position as we stride along but, unless we have been bullied into a showy military marching gait, we do not put much effort into this action. Even after a long cross-country walk, when our feet are aching and our leg muscles are exhausted, our gently moving arms still feel fresh and relaxed. It is only when we start to pull them right away from the body that they feel the strain of our endeavours.

The Arms Up posture is the hardest to hold for any length of time. It is the typical gesture of triumph and victory, much loved by politicians and sports stars. With their arms stretched aloft they greet their followers and celebrate their high status with a high posture. The raising of the arms makes them seem taller and stronger and also renders them more conspicuous at moments when they most want to be seen. They hold this position only for a matter of seconds, however. If they tried it for a matter of hours or even minutes, instead of seconds, they would soon find themselves suffering.

A totally different meaning applies when a thug with a gun commands 'Hands up!' Here too the arms are raised, but in defeat rather than victory. There is a subtle difference, however, in the angle of the arms in the two cases. In the victorious posture, the arms are usually kept straight or, if they are bent, are angled slightly forwards. In the gun-threatened posture, the arms are typically bent slightly at the elbow and are in the vertical plane rather than angled forward. The essence of the defeated posture is that it should indicate limp, ineffectual arms and hands, removed as far as possible from the body, where a weapon of some kind might be hidden.

The Arms Spread posture is the long-distance embrace-invitation gesture. A woman greeting an old friend who is still some paces away may fling wide her arms and hold them there until they can be wrapped around the friend's shoulders in an emotional hug. This same posture is seen after a circus performer has completed a difficult trick. She will fling wide her arms and the audience will immediately respond by applauding. The performer has invited an embrace and the audience has obliged with the only sort of gesture they can manage from their seats in the auditorium. Their action of clapping the hands together is in origin a highly modified form of 'vacuum embrace', in which the feel of the hug has been converted into the *sound* of a symbolic hug.

The Arms Forward position is more complicated. This can either signal rejection, if the palms are pushing forward; or aggression, if the fists are clenched; or begging, if the palms are held face-up. Like the spread arms it can also be an embrace-invitation gesture, and it can transmit a whole variety of other signals, according to the way the hands are being used.

Specialized arm signals include the different forms of waving, beckoning and saluting, each with its own particular flavour. When an important female figure gestures from a balcony her arm movements can be seen from a great distance. Their exact shape and style indicate something of her mood. The Queen's royal wave is the modestly elevated, rather limp gesture of passive power. By contrast, the clenched-fist salute of the female rebel leader is the fierce sign of active revolutionary power. The Nazi *Heil* was the stiff flat-handed action of rigid loyalty. The Military Salute – bent elbow,

124

and hand to headgear – is the stylized intention movement of raising the vizor or removing the helmet, an appeasing act meant to cancel the hostile signal otherwise transmitted by the gesturer's appearance. And so on. Arms are used for gesturing whenever we need a long-distance signal of a cruder kind than those transmitted by fingers and facial expressions. In this role, female arms act as invaluable body-flags.

In person-to-person contact the arm region is often a focus of friendly non-sexual actions. If we help an elderly stranger across the street we take their arm to support them. If we guide someone through a door we gently direct them by their elbow. If we wish to attract a stranger's attention we tap them on the arm. If in any of the contexts we touched waist, chest, or head our actions would immediately come under suspicion. The arms are the most neutral of body parts in this respect, as free of special intimate significance as any body zone can be. Friends may link arms when walking together, regardless of gender; but if any other kind of contact was made while walking it would promptly signal intimacy of a special order.

Arms have often been tattooed; but the most common form of adornment has always been the bracelet. Bracelets have nearly always been worn by females and it has been suggested that the custom originated as a way of exaggerating the gender signal of the slender feminine arm, the fine bracelets emphasizing the thinness of the arm diameter inside them. A rival suggestion is that they appeal to males because they are symbolic manacles, suggesting the enslavement of the women by their men.

13. The Hands

Female hands are superior to male hands in one important respect – they are more flexible. They may be smaller than the beefy paws of adult males and they may lack the immense gripping power of the male hand, but they have far greater finesse when it comes to the delicate handling of small objects.

Where intricate finger-work is required, the female hand is unbeatable. To give one example: the piano keyboard was designed for the male hand, putting female players at an immediate disadvantage when spanning the notes. The result is that most of the great pianists are males. But if a slightly smaller keyboard was made, scaled down to suit the smaller size of the female hand, then the greater flexibility of the female fingers would mean that female pianists would easily outplay their male counterparts. In a similar way, rock-climbers report that female flexibility matches male strength, giving the two sexes equal potential when scaling difficult rock faces.

How did this come about? What is the evolutionary story of the female hands? What happened when, millions of years ago, our ancestors stood up on their hind legs and freed their front feet to develop in a new direction?

The key element in this story – the secret of success for the human hands – was the development of opposable thumbs. Freed from the task of locomotion, both on the ground and in the trees, the design of the hands could for the first time become solely manipulatory. This was one of the most important steps in the whole evolution of our species. The human species became dextrous – and crossed

126

a major threshold into a world where nothing was safe from its grasping fingers.

In a physical sense males are much more grasping than females. The average male hand has about twice the gripping power of the average female hand. This is one of the more pronounced gender gaps and reflects the great importance of strong hands to the primeval hunter. The typical male can exert a grip of about 40 kg (90 lb), and with special training can increase this to 54 kg (120 lb) or more. A vice-like grip was particularly useful for fashioning weapons and other early implements, for throwing objects with force and for other activities involving such actions as hammering, twisting, tearing, clinging and carrying. Even today, tasks that benefit from large powerful hands are still male dominated. There are few female carpenters.

The Power Grip is only one half of the manual success story, however. The other half is the equally important Precision Grip. Power is achieved by opposing the whole of the thumb against the whole of the fingers. Precision is achieved by opposing only the tips of the digits. In this action the female is superior to the male. The big male hands, although capable of great precision when compared with the short-thumbed hands of other species, cannot compete with the delicately agile, small-boned hands of the human female when it comes to finicky tasks. As a result, in the past, females have always excelled at such pursuits as sewing, knitting, weaving and the finer forms of decorative work. Before the pottery wheel was introduced women also dominated the important ancient art of ceramics, where nimble fingers were so important in shaping and decorating the vessels. Because pottery was the major art form of the prehistoric period it follows that during that long phase in the human story it was the females and not the males who were the important creative artists – a fact usually overlooked by archaeologists and art historians.

The situation has changed little today, although the exact nature of the tasks involved may have been updated. Take a look inside any factory that involves, say, the intricate assembly of the tiny working parts of electronic equipment, and you will see rooms full of agile female hands. Needle and thread may be less in evidence, but female manual dexterity remains an important commodity.

This difference in manual precision is not just a matter of women having lighter, thinner fingers. The female finger-joints are also more flexible, a feature thought to be influenced by hormonal factors. It has been argued that this was a special adaptation to the primeval feminine specialization of food-gathering, as opposed to hunting. Food-collecting, involving harvesting roots, picking seeds, nuts and berries, and selecting fruits, required the deft quick fingers of the fine-boned, loose-jointed female hand rather than the power-paws of the muscular male.

This physical division of labour, which has occurred during our evolution, has made men and women slightly less like one another – each better in certain ways. And of course the process of specialization never went too far. Women's hands remained reasonably strong and men's hands were capable of quite fine work. The strongest of women could always tear apart a piece of meat or (today) open a stubborn bottle-top better than the weakest men in any group; and sailors at sea have always proved adept with a needle and thread. There are even a few male harpists, with remarkably flexible fingers. But right from the earliest days of the Old Stone Age there has been a significant hand-bias – power for men and precision for women.

Of all parts of the human body the hands are perhaps the most active, yet we seldom hear of anyone complaining of 'tired hands'. As complex pieces of machinery they are superb. It has been estimated that during one lifetime the fingers bend and stretch at least 25 million times. Even newborn babies have remarkable strength in their fingers, and their hands are hardly ever still. As they lie in their cots their tiny fingers are flexing and twitching as if anticipating the pleasures of handlings to come. And later in life what handlings they prove to be: operating a keyboard at a hundred words a minute, playing concertos at breakneck speed, operating complex machinery, performing brain surgery, painting masterpieces, reading Braille with the fingertips, and even reciting poetry in sign language for the deaf. Compared with the Rolls Royce of the human hand, the other species do not even own a bicycle.

A pair of human hands contain no fewer than 54 bones. In each hand there are 14 digital bones, 5 palmar ones and 8 in the wrist.

The sensitivity of the hand to heat, pain and touch is fine-tuned and there are literally thousands of nerve-endings per square centimetre. The muscular strength of the hands and fingers comes not only from the musculature in the hand itself but also from more remote muscles in the forearm.

On the surface of the hands there are three kinds of lines – the Flexure Lines, the Tension Lines and the Papillary Ridges. The first of these, the Flexure Lines or 'skin hinges', are creases that reflect the movements of the hand. They vary slightly from individual to individual, a fact that has ensured a steady income for palmists down the centuries. Like those other confidence tricks, phrenology and astrology, palmistry lost ground rapidly in the twentieth century and is now at last no more than the fairground fun it deserves to be. Its one useful legacy is the naming of the various crease-lines in a way that can easily be remembered. The four main lines are the 'head line' and the 'heart line', running across the palm, and the 'life line' and the 'fate line', running around the base of the thumb. In apes, the head line and the heart line are one, but in humans the independence of the forefinger is such that it splits the line into two. Some people, however, still display the ancient condition – the single line or 'Simian Crease', right across the palm. It is present in roughly 1 person in 25.

The Tension Lines are the small wrinkles that increase with age and become permanent as the skin loses its elasticity. The tiny Papillary Ridges are the 'grip' lines that are the basis of fingerprints. Sweat makes these little ridges swell and become more elevated, helping the hand to grip objects firmly. The sweating action of the hands is unusual. When people are asleep the palmar sweat glands cease activity, no matter how hot they may be in bed. In fact, they do not respond at all to heat-increases like the sweat glands on other parts of the body. They only react to increased stress. If your palms are bone dry you are relaxed. As you become more and more anxious they get damper and damper, preparing themselves for the physical action which your system is anticipating. Unfortunately the human body evolved this reaction at a time when most stress *was* of a physical nature, but today our modern tensions are more likely to be psychological ones, with the result that you are left with damp sticky palms and nothing to grip. Palmar sweating is a relic of an ancient hunting past that most modern urbanites could do without.

During the famous Cuban missile crisis in the 1960s, when the Western world held its breath, fearing a nuclear war, all laboratory experiments into palmar sweating had to be abandoned temporarily. The general increase in stress raised the sweating rate to an extent that made it impossible to get a 'relaxed' reading from any of the subjects being tested. Such is the sensitivity of the human hand.

The fingerprints form three basic patterns: loops, which are very common; whorls, which are moderately common; and arches, which are rather rare. No two human digits have ever been found with identical fingerprint details. Despite popular opinion to the contrary even identical twins have different fingerprints. The use of prints for identifying individuals is centuries old. Over 2,200 years ago the Chinese were labelling their seals of authority with their personal finger marks. Since signatures can so easily be forged, it is surprising that we do not follow the ancient Chinese custom. The use of finger-printing in modern crime detection has become highly sophisticated, with the technique of 'ridge-counting' and attention to tiny line details bearing such names as 'lakes', 'islands', 'spun crossovers' and 'bifurcations'. There is no way that criminals can avoid this type of identification by trying to alter their prints. Even if they have them painfully worn away, they soon grow back again, and they do not alter with age.

Racial differences in fingerprints do exist and Caucasians, for example, have fewer whorls than Orientals, and more loops; but the differences are only slight.

There are three special colour qualities of the human hand that have aroused interest. When pale-skinned people become suntanned the backs of their hands become brown, but their palms refuse to darken. This special feature of the human palms is said to have evolved in connection with the need to keep hand gestures highly conspicuous. Even dark-skinned races have pale palms.

Anyone who has been snowballing will have discovered that after a period of time has passed the palms become bright red. This special response appears to be a mechanism to prevent damage to the sensitive palmar skin from chilling. The reaction to prolonged cold produces a dramatic increase in blood flow that warms the hands. This is a remarkable and complex response. The initial reaction of the hands to the cold snow is vasoconstriction, reducing the flow of blood to the surface. This is the usual reaction of the body as a whole. It prevents warm blood from dissipating the vital body heat from the surface. This response remains the same for the rest of the body, no matter how long the exposure to cold may last, but the hands operate independently in a special way. After about five minutes they switch from strong vasoconstriction to the exact opposite – strong vasodilation. The palm and finger blood vessels suddenly expand and the hand goes bright red. Then, after about another five minutes the process reverses itself. If the ungloved snow-baller were sufficiently stoical to keep going for an hour she would observe her hands turning from blue to red and back again every five minutes. This is an emergency protection system that we probably evolved back in the Ice Age, when frostbitten hands could easily spell disaster. By repeatedly warming the hand-surfaces for brief five-minute spells it prevents the prolonged chilling that can cause the real damage. By repeatedly allowing them to cool for five-minute intervals it conserves precious body heat.

One of the most extraordinary items of human handlore is the claim by certain holy people to be afflicted by *stigmata*. These are supposed to be spontaneously formed wounds on the palms of the hands, similar to those that Christ is assumed to have suffered on the cross. The vast majority of the 330 people on record who have displayed the bleeding stigmata have been Roman Catholics, including a number of nuns. Intriguingly, women suffering from the stigmata outnumber men showing these marks by a ratio of 7 to 1. The phenomenon has been known for over 700 years, from the thirteenth century to the present day.

The Church authorities have always been uneasy about such claims. It is not the wounds themselves that are in doubt but whether they were miraculously caused. In typical cases the wounds on the palms would suddenly start to bleed, then heal up, then bleed again. One stigmatic bled to a very tight schedule: between one o'clock and two o'clock every Friday afternoon, and then again between four o'clock and five o'clock.

The most likely explanation of the cause of the stigmata wounds, assuming there was no deliberate self-mutilation, is that they were cases of localized virus infection. Children using public swimming pools often catch verrucas – small virus-warts, which have to be removed surgically. Similar warts may also appear on the palms of the hands, although they are less common there. When they do develop, however, they are often scratched and start to bleed. The scratcher may not even recall touching them. After a while they heal, but the process is much slower than in an ordinary cut. Because of the virus, the healing is not perfect and sooner or later they start to bleed again, growing larger in the process. Surgery is needed to remove them permanently. It is easy to see how such a minor affliction could fire the imagination of a devout nun and become a miraculous re-enactment of Christ's suffering.

Unfortunately there is nearly always one fatal flaw – the stigmata appear in the centre of the palms of the hands, while in true cruci-fixions, the nails were driven into the wrists. Religious artists are to blame for this error, having shown nails through the centres of Christ's palms in their religious paintings and sculptures from the ninth century to the present day. It would seem that their error, which to them was no more than artistic licence, has then been slav-ishly and painfully copied by the would-be miracle-sufferers. Significantly, the few who have bled through their wrists have all appeared very recently, after it became known that this was the actual location of the crucifixion nails.

Turning now to the digits – each has its own qualities:

The first digit, the thumb, is without question the most impor-tant of the five digits, since it endows the hand with its grip. Its vital role has been recognized since medieval times, when compen-sation for the loss of a thumb was placed at more than four times that of the little finger. If a thumb is lost today, modern surgery can adjust the forefinger so that it works in opposition to the other fingers, to some extent restoring the hand's gripping action.

In Latin, the thumb was known as *Pollex*. In ancient times it was dedicated to Venus, presumably because of its phallic significance. In Islam it was dedicated to Mohammed. It has three key gesture meanings: it points direction, it delivers a phallic insult and it signifies that all is well.

The second digit, the forefinger, is the most independent and important of the four fingers. It is the one most used in opposition to the thumb for delicate precision actions. It is the finger that pulls the trigger, the finger that points the way, the finger that dials the phone, the finger that beckons, the finger that calls for attention, the finger that jabs an opponent in the ribs and the finger that presses the button.

It has been given many names. Because it indicates the way, it has been termed the index finger, the indicative finger, the demonstrative finger or the pointer. Because it fires a gun, it has been named the shooting finger or the trigger finger. It has also at various times been called the Napoleonic finger, the finger of ambition, the towcher (toucher), and the world finger. Its strangest title is that of the poison finger. In early days it was forbidden to use this finger for any sort of medication because it was believed to be venomous. This probably stems from the use of this digit in aggressive pointing and finger-jabbing, which gives it the symbolic role of a dagger or sword – something dangerous which may wound you, like the stabbing fangs of a snake.

Catholics dedicate the forefinger to the Holy Ghost, Islam to the Lady Fatima.

Despite its importance the forefinger is usually only the third-longest of the four fingers, being exceeded in most cases by both the middle finger and the ring finger. In 45 per cent of females, however, it is the second-longest finger, relegating the ring finger to third place. Surprisingly, this is true of only 22 per cent of males. Why there should be a significant gender difference in this respect is a mystery.

The third digit, the middle finger, the longest of the digits, acquired a whole range of names in ancient times, being known variously as the Medius, the Famosus, the Impudicus, the Infamis and the Obscenus. The reason for most of these names is that it was this finger that was employed in the most famous of Roman rude gestures. In this the other digits are bent tight while the stiffly erect middle finger is jerked vigorously upwards. The two bent digits on either side symbolize the testicles and the middle finger is the active phallus. This gesture has survived well during the 2,000 years since it was made in the streets of ancient Rome and in modern America is simply known as 'The Finger'. Its use by women, at least in the Western world, has increased greatly during recent years, the arrival of greater respect for sexual equality having brought with it greater gestural equality. In the past, obscene finger gestures have been almost exclusively male, but today's more assertive females are no longer shy about expressing themselves in this way.

In the more rarefied atmosphere of religion, the middle finger has quite different associations. In Catholicism it is the finger dedicated to Christ and salvation; in Islam it is dedicated to Ali, husband of Fatima.

The fourth digit, the ring finger, has been used in healing cere-monies for over 2,000 years. In the ancient rituals of the Aegean it was encased in a finger-stall of magnetic iron and used in 'magic medicine'. Later this idea was adopted by the Romans who called it the *digitus medicus* – the medical digit. They believed that there was a nerve running from the ring finger straight to the heart. They always used this finger when stirring mixtures because they thought that nothing poisonous could touch it without giving due warning to the heart. This superstition lasted for centuries, although the nerve to the heart sometimes becomes a vein and sometimes an artery. In medieval times the apothecaries were still religiously using this finger to stir their potions and insisted that all ointments should be rubbed on with it. The forefinger was avoided at all costs. For some, simply to stroke the ring finger over a wound was enough to heal it and it eventually became known as the healing finger or the leech finger. In parts of Europe it is still used today as the only finger suitable for scratching the skin.

If there is any practical value in this superstition it is that of all the digits the ring finger is the least used and therefore probably the cleanest. The reason for its comparative inactivity is that its musculature renders it the least independent of all the digits. If you make a fist with your hand and then try to straighten out and bend back each digit, one at a time, it is only the ring finger that refuses to straighten itself out fully – or does so with great difficulty. If the finger on either side of it is straightened at the same time there is no problem, but on its own it feels too weak to make the movement. This meant that it was the least likely of the digits to have been touching anything harmful and was therefore the safest for medical use. It must also have meant that it was difficult to use as an efficient stirrer without keeping the other fingers pressed back out of the way by the thumb.

It was because of this lack of independence that the finger also became known as the ring finger. The ancient custom of placing the wedding ring on the third finger of the left hand was based on the idea that the wife was committing herself to be less independent, like the symbolic digit. The use of the left hand was based on the idea that this was the weaker, submissive hand, appropriate for what was then considered to be the wife's subordinate role. It is only because these facts have been largely forgotten that we still use this finger today as a ritual part of the marriage service. If the true, sexist significance of the symbolism of the ring finger were more widely known, it would create an intriguing conflict for many a modern bride.

Because of its role as the ring finger, it was also known to the Romans as the *digitus annularis*. In Islam it was given to Hassan; and for Christians it was the 'Amen Finger', based on the fact the gestures of blessing were made with the Thumb (The Father), the Forefinger (The Son), and the Middle Finger (The Holy Ghost) followed by the Amen of the ring finger.

The fifth digit, the little finger, was known in Latin as the *minimus* or the *auricularis*, minimus because it is the smallest and auricularis because of its association with the ear. It is usually claimed that its title as the 'ear finger' is based on the fact that it is small enough to be useful in cleaning the ear, but this is probably a modern rationalization. In earlier days it was thought that by blocking the ears with the little fingers it was possible to increase the chances of a psychic experience, a prophetic vision, or some other supernatural event. Anyone who has attended a seance will probably have indulged in a modern survival of this superstition when joining hands in a circle. On such occasions the medium usually insists that it is the tips of the little fingers that are used to make contact with one's neighbours because this was the ancient way of forging a psychic link.

In America the popular name for this finger is the pinkie. The term was first used there by children in New York, but later spread to adults and to other cities. It is thought to have originated in Scotland, where children referred to anything small as a 'pinkie' and it is supposed to have been transported to the New World by Scottish settlers. However, the original name for New York was New Amsterdam and it may also be significant that the Dutch word for the little finger is *pinkje*. The children who employed the word often did so in a special rhyme that they chanted when making a solemn bargain. As they did so they interlocked their little fingers to make the bargain binding. This is yet another survival of the ancient psychic-link role of the little finger. In some European countries, when two people accidentally say the same word at precisely the same moment they shout 'Snap!' and then interlock their little fingers. While doing this they are both permitted to make a silent wish, which will come true if nothing is said until their fingers are released. Once again, this reflects the ancient belief in the psychic power of the little fingers and their ability to transmit supernatural forces. The reason for saying 'Snap' also has to do with the digits, because it is a verbal substitute for the action of snapping the fingers, another act with superstitious origins. It used to be thought that the loud noise made by the snap of the forefinger against the thumb would scare away evil spirits (which is why it is rude to snap your fingers for attention), and this was thought necessary when two people uttered the same word simultaneously.

In a completely non-magical context, the crooking of the little finger when drinking from a cup or a glass has long been thought of as the height of genteel affectation. In origin, nothing could be further from the truth. Early religious paintings often show the little finger crooked away from the other digits, even when the female figure in question is not drinking. It is claimed that this was a sign that the real life models for the religious figures were girls with an unusual degree of sexual independence. This belief that an 'independent' little finger symbolized sexual freedom was the basis of a new fashion started by the members of the women's movement at the end of the nineteenth century. They deliberately crooked their little fingers when drinking to display their support for the idea of equal rights in sexual matters. Slowly spreading as a fashionable finger posture, this action gradually shed its original meaning and eventually lost its sexual significance, becoming merely 'the thing to do' when in company. From there it was on its way to becoming a sign of polite gentility, ending up with almost the opposite meaning to its original one.

Used together, the five digits of the hand are capable of a huge range of gestures and signs, some deliberate and symbolic, others unconscious and expressive. Worldwide, women are less likely than men – even today – to employ the symbolic gestures, but they are more likely than men to make full use of the 'baton gestures' that accompany conversations and emphasize the words that are being spoken. Then, the female hand may be made into a grasping claw, a chopping edge, a stabbing point, a tight fist, or a spreading fan, according to the emotions of the moment. It is hard to recall, after a conversation is over, precisely what the fingers have been doing but, despite this, the message of the moving hands gets across to companions well enough at a subliminal level.

Adorning the female fingers has been popular for at least 6,000 years and probably much longer. By 2500 BC the goldsmiths of the Middle East had already reached an advanced stage in the manufacture of finger rings, and rings have never lost favour from that day to this. Originally they were worn for far more than mere decoration. It was believed that they also had important protective powers, bringing the wearer good fortune, protecting her from evil spirits, and offering her great wealth and even immortality (because the ring has no beginning and no ending).

A special advantage of ancient finger rings, and one that we would not think of today, is that, before there were efficient mirrors, rings were more appreciated than any kind of head or neck adornment, for the simple reason that they could be seen clearly by their wearers.

At a later date, another special benefit that they bestowed on some of their female wearers was the ability to dispose of unwanted males, it being simple enough to add ornate decorations to finger rings containing small secret chambers filled with lethal poisons.

The skin of the female hands has received comparatively little decorative attention, with the attractive exception of the application of henna patterns. These have been popular in North Africa, the Middle East and parts of Asia for centuries, and are an important part of the wedding ceremonies. Henna is an orange-red stain made from the powdered leaves of a small shrub. The intricate patterns that are laboriously painted on to the bride-to-be's hands are thought to protect her from the Evil Eye, a malevolent spirit that is always attracted to happy occasions with the intention of spoiling them. Henna is believed to possess a 'virtue' that purifies the bride from all earthy taints and renders her immune to the assaults of the Devil and his agents.

The night before the wedding, the bride, surrounded by her close female friends, must submit herself to the attentions of a special female artist called a *hennaria*. This artist will spend hours applying the traditional patterns, after which she bandages the bride's hands and places them inside two embroidered bags to ensure that they dry without smudging. This event is called the henna night – from which we seem to have taken our phrase 'the hen party'. For the wedding, the decorated hands are unwrapped and the beautiful patterns are displayed. The patterns normally last about four weeks, after which they can be allowed to fade or may be renewed. Today, a minor fashion for purely decorative henna-hands has appeared in both Europe and America, but the inconvenience of the designs has prevented it from become a mainstream form of female adornment.

In its undecorated condition, the skin on the backs of the female hands can pose a serious problem for the older woman. If she has rejuvenated her face by 'firming creams' or, more drastically, by a face-lift, making her look 20 years younger, she may find to her dismay that her true age is given away by the scrawny, wrinkly appearance of her hands. In earlier times, she could have worn an ornamental pair of gloves, but that useful form of cover-up is no longer fashionable. Tougher measures are needed to bring her hands into line with her younger look, and today a bewildering plethora of expensive procedures is available to her, including such dubious delights as micro-dermabrasion, acid peel, ultrasound, vitamin infusion, oxygen boosting, hot wax, acid multi-injections and laser treatment. The most radical procedure on offer is the manual equivalent of a face-lift – the hand-lift. This involves taking fat from the thighs and injecting it into the backs of the hands. This plumps them up and makes them look remarkably young again, but the treatment has to be repeated several times and even then lasts only for a year or so.

Finally, there are the fingernails – dead tissue growing from a living base at a rate of one millimetre every ten days – about four times as fast as toenails. This rate of growth means that, uncut, the fingernails would become about one centimetre longer in 100 days. In primeval times this would have been counteracted by natural wear and tear. In modern times regular cutting or filing is required to keep the nails at a convenient length.

Many women in different epochs and in different cultures have ignored convenience and have allowed their fingernails to grow long as a sign that they are not required to undertake any form of manual labour. This high-status display is enhanced by the application of brightly coloured varnish to the nails to draw attention to the fact that *these* are hands that never have to toil. In ancient China, noble women grew their nails long for this reason and painted them gold. Later, because this made their ordinary hand movements so cumbersome, they restricted the display to the little fingers only, clipping the others much shorter. Another solution was to have shorter nails for everyday use and then to clip on wildly exaggerated false nails for special occasions. Both these practices are found today in Europe. Many women use false nails that they glue on for social events and then remove for work.

With some eccentric individuals the fingernails have been allowed to grow to astonishing lengths, making ordinary, day-to-day hand-actions extremely difficult. Telephone numbers, for example, have to be dialled with the knuckles. One Dallas nail-fanatic boasted a total of 380 cm (150 inches) of nail on her hands, her most impressive one measuring 71 cm (28 inches). It took her eight to ten hours to paint her amazing talons. After caring for them for 24 years, she finally decided to have them trimmed. Once they were removed, her first great pleasures were going to be to scratch herself, and then to give someone a hug.

When they grow very long, female nails do not grow straight, they start to curve, and this in itself can cause problems. A woman in Augusta, Georgia, who had been charged with a misdemeanour, routinely had to have her fingerprints taken. When the police attempted to do this, they found that her 15cm long (6 inches), curved talons made it impossible and ordered her to cut her nails. She refused and, to save her spectacular hands from abbreviation, was forced to spend four nights in jail, while police arranged for a special way of obtaining her fingerprint impressions.

Long female nails can easily become weapons of male destruction. A jilted Connecticut woman was so outraged at finding her partner in bed with another woman, that she employed her manicured hands to take her revenge. The injured man required 24 stitches to close the wound in his scrotum.

In recent years, the fashion for long painted fingernails has been amplified by the introduction of 'nail art', in which fancy patterns are added to the surface of the nails. Astonishingly, there are now 61,400 websites on the internet dealing with the subject of 'nail art' and there is even a *Nail Art Encyclopedia* available for those who wish to take this subject seriously. There is themed nail art, free-hand nail art, encased nail art; and there are stick-on rhinestones, acrylic nail sculptures, hologram tips, and dangling nail jewellery for pierced nails. The list is endless.

Many women have found this elaborate nail art too exotic and overstated, and have turned instead to a newer style, the French Manicure, that gives the appearance of a natural-looking nail, but with emphasized white tips. Alternatively, nails have been cut short and painted with near-black polish, as the fashion world continues to ring the changes.

It is easy to smile at these cultural exaggerations and decorative embellishments of female fingernails, but a tradition that has lasted for over 6,000 years, in one form or another, is hardly likely to disappear overnight. Providing that the modifications do not interfere with the mobility and flexibility of the female hands, there is no harm in them. And even when they do hamper female hand actions, their social impact may be so rewarding for their owners that it compensates for the loss of manual dexterity. (Just so long as they do not suffer the fate of a Massachusetts woman whose long fingernail became wedged in the slot of a parking meter and who required the police and the local fire brigade to come to her rescue and release her.)

14. THE BREASTS

The female breasts have received more erotic attention from males that any other part of the body. Focusing such attention directly on the genitals is too extreme; on other parts of the anatomy not extreme enough. The breasts are the perfect happy medium – a taboo zone, but one that is not too shocking.

As a result, breasts have attracted an amazing range of euphemisms. No fewer than 74 colourful titles have been ascribed to them over the centuries, including such exotica as Big Brown Eyes, Brace and Bits, Cats and Kitties, Charlie Wheeler, Cupid's Kettledrums, Golden Apples, Mae West, Moons of Paradise and Twin Globes. Less flowery terms include bosom (tenth century), paps (fourteenth century), duckys (sixteenth century), bubbies or tits (seventeenth century), bust or diddies (eighteenth century), dugs or titties (nineteenth century), boobs, bristols, gazungas, hooters, jugs, knockers, mammaries or melons (twentieth century).

The breasts of the human female have two biological functions, one parental and the other sexual. Parentally they act as gigantic sweat glands producing the modified sweat we call milk. The glandular tissues that produce the milk become enlarged during pregnancy, making the breasts slightly bigger than usual. The blood vessels serving these tissues become much more conspicuous on the breast surfaces. As the milk forms it passes along ducts towards special storage spaces called sinuses. These are positioned in the centre of the breast behind the dark-brown areolar patches that surround the nipples. From these sinuses there are some 15 to 20 tubes, the lactiferous ducts, leading to each nipple.

When a baby sucks it takes the whole of the areolar patch and the nipple into its mouth, squeezing the brown skin with its gums and squirting the milk out of the nipple. If it takes only the nipple into its mouth it has a problem, because squeezing the nipple alone does not produce the desired milk. It may respond to this frustration by chewing on the nipple, which does little good either to mother or offspring. An inexperienced mother soon finds that she can avoid the pain caused by these hungry attentions by squeezing more of her breast into the baby's mouth.

The areolar patch surrounding the nipple is an intriguing anatomical detail of the human species. In virgin females and those who have yet to become mothers it is a pinkish colour, but during pregnancy it changes. About two months after conception it starts to grow larger and becomes much darker. By the time lactation has started it is usually a darkish brown colour and later, when the baby is weaned, it never quite returns to its original virginal pink. In function these areolar patches appear to be protective. They are full of specialized glands that secrete a fatty substance. To the naked eye the glands look like 'goose pimples' on the pigmented skin. During the breast-feeding phase they become much enlarged, and are then called Montgomery's tubercles. Their secretions help to protect the skin of the nipple and its surrounding skin – a form of biological 'skin care' much needed by the lovingly abused surface of the breast.

The milk produced by the female breasts contains proteins, carbohydrate, fat, cholesterol, calcium, phosphorus, potassium, sodium, magnesium, iron and vitamins. It also contains various antibodies that may make the infant more disease-resistant. Cow's milk is a fair substitute for mother's milk, but its phosphorus level is rather high and may interfere with the intake of calcium and magnesium by human infants. Also, some babies may have allergic reactions to bovine proteins. Wisely, more mothers are breast-feeding nowadays – and there is the added bonus that it forms a much stronger bond of love between mother and baby.

While mother's milk is ideal for a growing baby it has to be said that the shape of her breast is far from perfect for the task of breast-feeding. The teat of a milk bottle is much more suitably shaped for delivering milk into the infant mouth than the real nipple on the mother's breast. If this appears to be an evolutionary flaw, it must be remembered that the female breast has a dual role – parental and sexual – and it is the sexual factor that causes the problem here. To understand why this should be, it helps to cast a sidelong glance at the breasts of our close relatives, the monkeys and apes.

In all other primate species the females are flat-chested when not lactating. When they are breast-feeding, the region around the nipples becomes somewhat swollen with milk, but even then it is rare to find anything approaching the hemispherical shape of the human female breast. In the case of those few that do approach the human shape when they have a particularly generous milk supply the swellings disappear after lactation is complete. The 'breasts' of monkeys and apes are purely parental.

The breasts of the human female are different. Although they increase in size slightly when full of milk they remain protuberant and firmly shaped throughout the period of young adulthood regardless of parental consideration. Even a nun has protuberant breasts though they remain unused throughout her lifetime.

An examination of the anatomy of the breast reveals that most of its bulk is made up of fat tissue, while only a small part is glandular tissue concerned with milk production. The rounded shape of the breast, created by this fat tissue, therefore requires a separate explanation, one that goes beyond milk production. Although it is clear to a biologist that this explanation has to do with sexual signalling, some women have objected to this interpretation. They find offensive the idea that some aspects of the female body might have evolved their present form to appeal to the human male. Ignoring the fact that physical sex appeal was involved in their own conception, they insist that the female breast must be totally parental in function and apply their ingenuity to finding non-sexual explanations for the evolution of breast roundedness. They have put forward seven suggestions:

The fat tissue cushions the milk glands. This may be true during lactation, but does not explain the persistent roundedness at other times. And it fails to explain why other primates have not needed this aid.

The fat tissue keeps the milk warm. Again, this is needed only during lactation.

The roundedness of the breast makes it more comfortable to feed the baby. This is simply not true. One only has to think of milk bottle design.

The roundedness acts as a visual signal telling males that the owner of a large breast will be a good nursing mother. Again, this is not true. Small-breasted women can breast-feed more easily than big-busted ones.

The fat tissue acts as a valuable form of fat storage for when food is scarce. Yes, it does, but why concentrate this storage on the chest, where large protruding breasts make fast running more difficult? The female body has a generous layer of fat over most of its surface and this wide-spreading of its fat storage is the most efficient way of insuring against the hazards of temporary famine. Furthermore, the breast fat represents only 4 per cent of the female body-fat total, and it is the least variable fat when it comes to weight loss.

The fat tissue compensates for the absence of a coat of maternal fur, to which the baby can cling when feeding. This is not true. As any mother knows, the human baby has to be held to the breast and, in any case, a smooth, large hemisphere of flesh hardly helps to make the nipple more accessible.

The hemispherical shape of the breast is, according to one author, 'non-functional to the point of being counter-functional'. When all the other parental explanations are seen to be faulty, this is the last ditch stand of those who refuse to accept that the female breast shape is sexual.

The inescapable conclusion is that the hemispherical shape of the breasts is not a parental development. It is concerned instead with sexual signalling. This means that suggestions that men's interest in women's breasts is 'infantile' or 'regressive' are unfounded. The male responding to the prominent breasts of a virgin or non-lactating

145

female is reacting to a primeval sex signal of the human species.

The origin of paired hemispheres as a human female sexual signal is not hard to find. The females of all other primates display their sexual signals backwards from the rump region as they walk about on all fours. Their sexual swellings are key stimuli that excite their males. The rump signals of a human female consist of unique paired hemispheres, the buttocks. These can act as powerful erotic signals when she is seen from behind, but she does not go around on all fours like other species, with her frontal region hidden from view. She stands upright and is encountered frontally in most social contexts. When she stands face-to-face with a male her rump-signals are concealed from view, but the evolution of a pair of mimic-buttocks on her chest enables her to continue to transmit the primeval sexual signal without turning her back on her companion.

So important was this sexual element in breast development that it actually began to interfere with the primary parental function. The breasts became so bulgingly hemispherical in their efforts to mimic the buttocks that they made it difficult for a baby to get at the nipples. In other species the female nipples are elongated and the baby monkey or ape has no difficulty whatever in taking the long teat into its mouth and sucking away at the milk supply. But the human infant of a well-rounded mother may be almost suffocated by the great curve of flesh that surrounds the rather modest human nipple. Such mothers have to take precautions that would never be needed in any other species. Dr Spock advises: 'At times you may need to put a finger on the breast to give him breathing space for his nose.' Another baby book comments: 'It may surprise you that he takes the brown area around the nipple in his mouth as well. All you need to do is to make sure that he can breathe. In his eagerness he may obstruct his nostrils with breast tissue or with his own upper lip.' Cautions such as these leave no doubt about the dual role of the human breasts.

Women who have rather small breasts often worry that they will not be able to breast-feed. Ironically they may well be able to breast-feed more efficiently than their well-rounded friends. This is because they have less of the fat tissue that gives the breasts their sexual hemispherical shape but which has little to do with the milk supply.

Once they become pregnant their glandular tissue will increase in size, as with all expectant mothers, but they will not have such bulbous breasts as heavier women and their babies will find sucking much easier and less suffocating.

In their sexual role the female breasts operate first as visual stimuli and then as tactile ones. Even at a great distance they are normally sufficient to distinguish the silhouette of the adult female from that of the male. At a closer range this crude gender signal gives way to a more subtle age-indicator. The shape of the breast changes gradually from the age of puberty to old age. This slow alteration in mammary profile can be simplified into the 'seven ages of the female breast' as follows:

The nipple breast of childhood. Only the nipple is elevated in this pre-pubertal stage.

The breast-bud of puberty. At the very start of the reproductive phase, when menstruation begins and the genitals start to sprout pubic hair, the region around the nipple starts to swell.

The pointed breast of adolescence. As the teenage years pass there is a further slight increase in breast size. At this stage both the nipple and the areolar patch project above the breast, creating a more pointed conical shape.

The firm breast of young adulthood. The ideal physical age for the human animal is twenty-five. This is the stage at which the body is at its peak of condition and all growth processes have been completed. During the twenties the female breast fills out to its most rounded hemispherical condition. Although it is larger its weight has not yet started to make it droop.

The full breast of motherhood. With maternity and the sudden additional bulk of enlarged glandular tissue, the milk-laden breast balloons out and starts to turn downwards on the chest. The lower margin of the breast overlaps the chest skin to create a hidden fold.

The sagging breast of middle age. As the reproductive phase of adulthood nears its end, the breasts hang further down on the chest even though they have lost the fullness of the lactation stage.

The pendulous breast of old age. With advancing old age the general shrinkage of the body leads to a flattening of the breasts that remain hanging down on the chest but with increasingly wrinkled skin.

There are many variations in these typical stages in breast aging. In thinner women the process tends to slow down to some extent, while in fatter ones it speeds up. Cosmetic surgery can prop up breasts and artificially extend the firmness of the young adult stage. Costume supports such as corsets and bras can give the same impression, providing the breasts are not directly visible. In a variety of ways over the years women have sought to prolong the impression of firmly protruding hemispherical breasts in order to extend the period over which they are able to transmit the primeval female chest signal of the human species.

Sometimes the mood of society has demanded that the sexuality of the female bust be suppressed. Puritans achieved this by forcing young women to wear tight bodices that flattened the breasts and gave an innocent childlike contour even to mature adults. In seventeenth-century Spain young ladies suffered even greater indignity, having their swelling breasts flattened by lead plates that were pressed tightly on to their chests in an attempt to prevent nature from taking its protuberant course. Such cruel impositions only serve to underline the intense sexual significance of the hemispherical shape of the breast. For society to go to such lengths to negate it, it must indeed be powerful.

Happily most societies have been prepared to accept the covering rather than the crushing of the breast as a sufficient expression of modesty. In such instances the mere removal of this covering has acted as a major erotic stimulus. This has been exploited by artists and photographers in many different ways. For the artists it is easy enough to produce the perfect breast; they can invent any breast shape they like within reason. If they stray too far from nature the

HAIR

ABOVE: *In a young female each of the 100,000 head hairs grows seven inches a year and lasts for at least six years – providing a display that must have given primeval woman, with untrimmed hair, a unique appearance.*

RIGHT: *Although only a small proportion of the 3,000 million women in the world are naturally blonde, many more lighten their hair to adopt this colour, giving it a softer, more juvenile look. Blonde hair is often worn loose and long, symbolizing lack of restraint.*

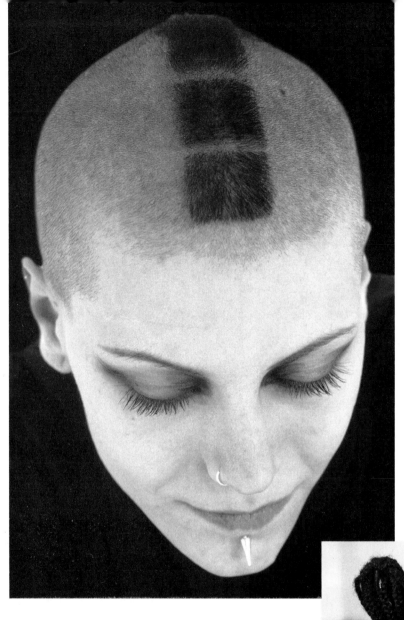

ABOVE: *Extreme rebellion calls for something outrageous – in this case the carefully shaved head of the punk female. In ancient times, a shaven female head was associated with slavery, punishment or mourning.*

RIGHT: *The modern braided look is the latest in a long line of elaborate, time-consuming hair decorations, an interlacing tradition that stretches back into the Old Stone Age, 21,000 years ago.*

HAIR

LEFT: *In the 1960s Vidal Sassoon reintroduced the Short Look that had been briefly popular during the Flapper epoch of the 1920s. It was a tomboy style that emphasized the active, carefree mood of the period.*

RIGHT: *The Big Hair Look of the 1980s – a height-increasing strategy. Unlike the tall wigs of earlier centuries, this employed the owner's real hair that was groomed and sprayed into a rigid tower.*

EYEBROWS

ABOVE: *As part of the feminist rebellion, certain women deliberately wore heavier eyebrows, sometimes called 'caterpillar brows', as a way of displaying a more assertive, more fiercely determined facial expression, and at the same time rejecting elaborate grooming.*

LEFT: *The most extreme form of female eyebrow-display is that of the Latin American artist Frida Kahlo with her famous unibrow or monobrow.*

Because female eyebrows are less bushy than those of the male, they have often been trimmed thinner to make them super-female, as here with Mae West (ABOVE).

In the West today a few women display pierced eyebrows, but in tribal societies brow scarification (ABOVE RIGHT) has been commonplace for centuries.

The raised eyebrows of surprise (RIGHT) and the cocked eyebrows of the quizzical (BELOW) are just two of the six eyebrow signals that indicate changing moods.

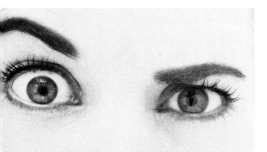

EARS

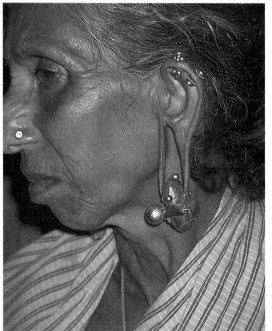

Female ears have been brutalized in many cultures all over the world, with the long-suffering earlobes pierced, stretched and decorated in a hundred different ways. Originally this was done to increase the wisdom of the woman, or to protect her from misfortune, but today it is largely aesthetic. The stretching of the lobes is carried to extremes in some African (ABOVE RIGHT) and Asiatic (LEFT) tribes. And in the West today some women subject themselves to multiple piercings (ABOVE LEFT).

RIGHT: *Queen Cleopatra was renowned for her experimental eye make-up colour combinations, especially blue and green. When Elizabeth Taylor appeared as Cleopatra in 1961, wearing elaborate, heavy eye make-up, she started a trend that was to influence fashion for several years to come.*

BELOW RIGHT: *The one-eyed blink that we call the wink is a light-hearted signal of collusion that some women find difficult to perform in a casual way.*

BELOW: *The epicanthic eye-fold of Oriental woman is today often surgically removed to give its owner a more Western appearance.*

EYES

EYES

ABOVE: *Each human female possesses a total of about 400 eyelashes. Artificial extensions have often been used to exaggerate the size of the eyes and to increase the impact of the fluttering eyelash signal.*

BELOW: *Because eye contact is the key element in all social interaction, the appearance of the eyes cannot be missed. As a result, elaborate eye make-up remains a favourite of high fashion today.*

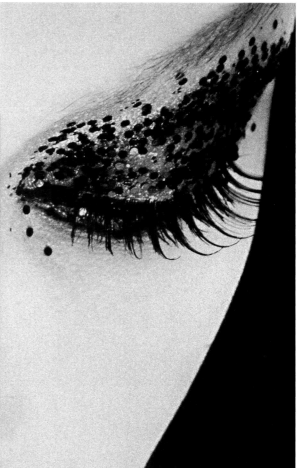

As a site for attaching ornaments, the nose has always been less popular than the ears because, unless the ornaments are rather small (ABOVE) *they hang down over the mouth* (RIGHT) *and interfere with the actions of eating, drinking and speaking. Despite this, some tribal women traditionally have the nasal septum pierced and display valuable nose ornaments on special occasions.*

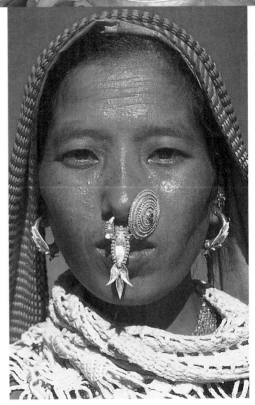

NOSE

CHEEKS

The smooth beauty of the female cheek is usually left unmarked and, if it is adorned, is merely powdered or painted to look even smoother. But from time to time it receives more exaggerated forms of attention, including smearing with face paint (in Madagascar, ABOVE), decorating with tribal patterns (in Kenya, RIGHT), whitening (in Japan, BELOW LEFT), or even being adorned with World Cup football designs (Korea, BELOW RIGHT).

LEFT: *The 'cupid's bow' lip-fashion displays an exaggerated dip in the middle of the upper lip and a deep lower lip, giving the woman a more baby-faced appearance. This was popular in the 1920s, with Hollywood actresses, like the aptly named Clara Bow, seen here, who made it a special feature of her facial make-up. In the Far East (BELOW), Japanese Geishas adopt a similar style, but with even greater emphasis on the deep lower lip.*

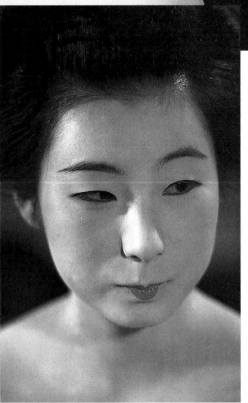

LIPS

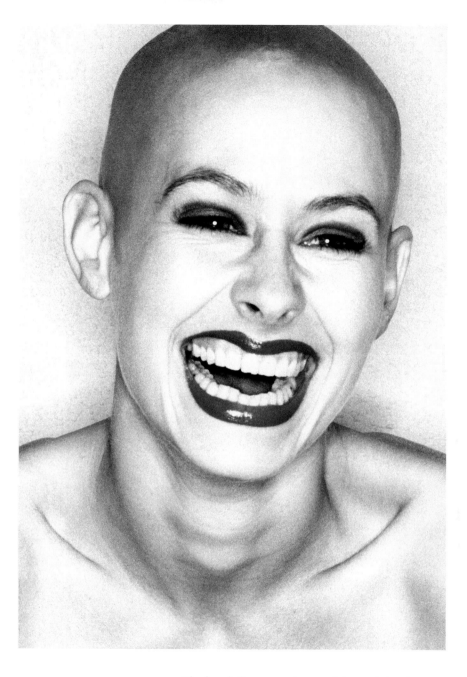

LIPS

The female lips transmit powerful sexual signals because they mimic the colour, shape and texture of the labia. The labia become redder during sexual arousal, which explains why red has always been the dominant colour for lipstick, regardless of attempts by cosmetic companies to introduce new colour variations.

Tribal lip decorations include an insertion of massive lip plates (in Africa, ABOVE LEFT) *and lip tattooing* (among the Japanese Ainu, ABOVE). *In some African tribes, the worth of a woman is measured by the size of her lip plate.* LEFT: *In Western society a new fashion for enlarged lips has led some women to undergo painful cosmetic enhancement, creating what has been cruelly called the 'trout pout'.*

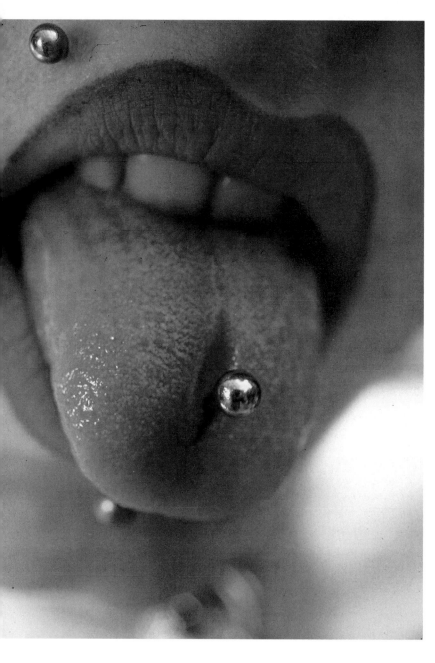

MOUTH

Because the human mouth is so busy talking, singing, chewing, licking, tasting and swallowing, it has rarely been mutilated for decorative reasons. A recent exception to this is the metal tongue-stud favoured by some young Western women. Although it creates a slightly lisping voice, it appeals to the more rebellious spirits because it upsets the older generation.

NECK

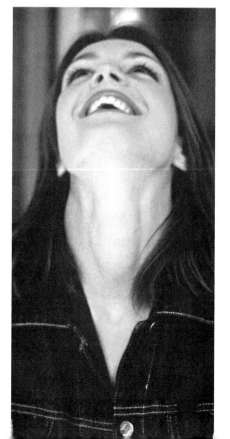

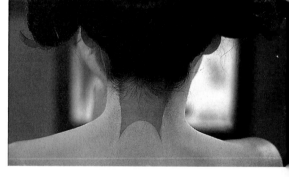

LEFT: *The female neck is longer and more slender than that of the male. This makes it the perfect site for decorative necklaces (as with this Karo woman,* ABOVE RIGHT), *and the Paduang woman from Burma* (ABOVE LEFT). ABOVE: *In Japan, the back of the neck is a site of a great deal of erotic attention; Geishas go to great lengths to decorate and display this part of their bodies.*

ABOVE: *The smoothly rounded contour of the female shoulder transmits an erotic signal, especially when emphasized by the 'chin on naked shoulder' pose. The hemispherical shape of the raised shoulder acts as a body-echo of the female breast or buttock.*

RIGHT: *In the 1980s, female 'power dressing' included the introduction of large shoulder-pads, creating a squared, masculine appearance. This strategy gave women a more aggressive look that had not been favoured since the tough wartime years of the 1940s.*

SHOULDERS

ARMS

ABOVE: *Arm bracelets have been popular for thousands of years.*

ABOVE LEFT: *When actress Julia Roberts displayed hairy armpits at a recent film premiere, it created a major stir in the world of cosmetics.*

LEFT: *Female bodybuilders develop massively muscular arms that are unattractive to most males, not merely because the arms appear so masculine, but also because the huge effort required to develop these muscles suggests an extreme form of narcissism.*

ABOVE: *The female hand is much more flexible than that of the male. This is perfectly demonstrated by oriental dancers, who emphasize this flexibility by adding elongated finger-stalls that exaggerate their hand gestures.*

LEFT: *Some modern women prefer to grow their nails unusually long, so that their hands look like savage claws. The presence of long nails also makes it clear that these women are incapable of carrying out any form of manual labour, and must therefore enjoy high status.*

HANDS

ABOVE: *Because they are kept so busy, hands are rarely decorated, but an exception to this is the application of henna patterns. In many parts of North Africa, the Middle East and Asia, the bride is decorated in this way on the eve of her wedding, at a special 'henna party' (which gives us our 'hen party'). The complicated patterns are thought to protect her from the Evil Eye and they last for several weeks.*

FAR LEFT ABOVE: *The bikini is not a new invention, as illustrated by these young women of the third century A.D.*
FAR LEFT: *The services of a structural engineer were enlisted to develop a specially cantilevered brassiere for Jane Russell.*

ABOVE: *In the 1950s, actresses often wore conical bras that gave the impression they had aggressively pointed breasts shaped like warheads. This strange fashion resurfaced briefly in a Madonna video in 1994 (ABOVE LEFT).*

LEFT: *Over one million women have had their breasts artificially enlarged by implants in recent years.*

WAIST

ABOVE: *The female waist:hip ratio differs significantly from that of the male – 7:10, compared with 9:10. To exaggerate this difference, a super-feminine silhouette became the goal of high fashion in earlier centuries, achieved by corsets so tight that brutal measures had to be taken to lace them. Even today, tight corsets still appear from time to time as an erotic novelty (LEFT).*

ABOVE: *The swaying body movements of the female Hawaiian dancer emphasizes the shape of her wide hips, focusing the attention of watchers on the flexibility of her childbearing pelvic region.*

RIGHT: *The arms akimbo posture – hands on hips – is an essentially unfriendly posture, with the protruding elbows saying 'keep away from me'. It is an anti-embracing posture seen when a figure is demanding more space.*

HIPS

BELLY

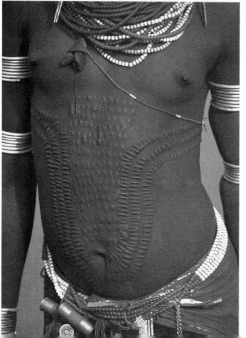

ABOVE: *The recent fashion among young women in the West for exposed bellies has encouraged the spread of decorative navel-piercing. Rare in the past because it went unnoticed, the pierced navel has now become a common sight. Among tribal people, such as the Karo of Ethiopia (Left), where the whole of the belly region is exposed to view, more extensive decoration is found in the form of patterned scarification.*

BACK

ABOVE: *Back ornaments are comparatively rare, except where tribal women leave this part of their bodies fully exposed to view.*

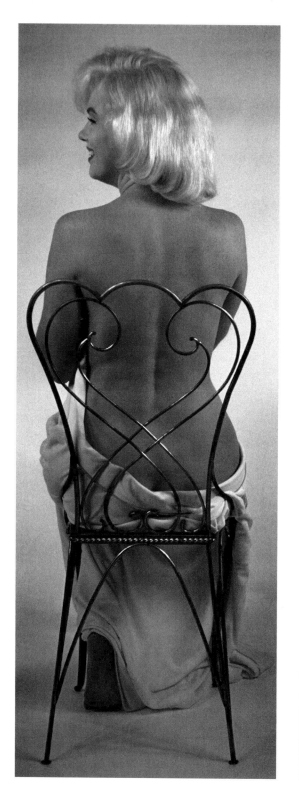

BACK

LEFT: *In the West the female back was first bared as a fashion statement in 1932 and since then has appeared from time to time as a politely daring exposure of a large expanse of naked flesh.*

BELOW: *The back provides the best 'body canvas' for the tattooist's art and there are some dramatic examples to be found on female backs today.*

PUBIC HAIR

ABOVE LEFT: *The display of female pubic hair has always been a powerful taboo in urban society and its accidental exposure has led to various forms of depilation. However, even these careful trimmings do not always have the desired effect when young nightclubbers are caught unawares by the camera* (LEFT).

ABOVE: *Women at the Tapati festival in Hanga Roa on Easter Island solve the problem of 'modest nudity' by resorting to the old custom of wearing a pubic wig, or merkin.*

ABOVE: *The protruding female buttocks transmit a powerful gender signal, both in sportswomen like Serena Williams and performers like Kylie Minogue (ABOVE LEFT) and Jennifer Lopez (LEFT). Their protrusion is emphasized by the backward rotation of the female pelvis.*

BUTTOCKS

ABOVE: *The well-known heart symbol, with the cleft in its upper surface, looks very little like the real heart, but does have an uncanny resemblance to female buttocks seen from behind.*

LEGS

As young girls reach sexual maturity their legs become longer in relation to the rest of their bodies. It follows that exceptionally long legs have greater sex appeal (OPPOSITE PAGE, TOP RIGHT), Wren Scott's legs, the longest in the world (FAR LEFT), measure 49 inches (124 cm.)

Part of the fascination of legs is that they focus attention on the point where they meet, and this was never more obvious than in Sharon Stone's infamous leg-crossing scene (OPPOSITE PAGE, BELOW).

TOP: For a thousand years girls of high status in China were forced to have their feet bound tightly until they became permanently crushed.

BELOW: Some high-status women have demonstrated their power and wealth by the size of their shoe collections. Imelda Marcos reputedly owned more than 3,000 pairs.

FEET

FEET

RIGHT: *Fashion designers seem to care little about the practical demands of female locomotion. Some shoes become so outlandishly clumsy that even experienced models lose their balance.*

BELOW: *The most extraordinary example of high-status female footwear went on display in 2003 – a pair of stiletto-heeled shoes by designer Stuart Weitzman, boasting 642 rubies set in specially spun platinum, inspired by the magical ruby slippers in* The Wizard of Oz. *They were put on sale at a price of one million pounds (approximately 1.5 million dollars).*

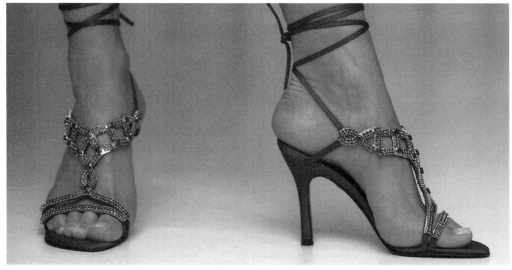

primeval signal becomes distorted and its impact is lost. But if the basic hemispherical shape is made slightly more hemispherical than usual it is possible to create a super-breast which is perhaps even more stimulating that the real thing.

The photographer has a more difficult task. Confined to real breasts the only hope is to improve their shape by special lighting or by arranging the models in postures that enhance the hemispherical signals. Of course the photographer can ensure that the models are both metaphorically and literally at their peak of breast development. To capture a super-breast requires a model whose adolescent breasts have reached maximum development just before their increasing weight starts to make them droop. There is a conflict of forces here because the increase in size that gives the full hemispherical shape inevitably also adds to the weight and starts to pull the breast downwards. There is only one point in a female's life when her breasts have maximum protrusion with minimum droop and that is the moment when the camera shutter must click to produce the most erotic images.

It is interesting that expert photographers who work for glossy magazines specializing in erotic pictures find that there is only one kind of girl with the super-breasts they seek. Her age is slightly younger than one might expect, namely the late teens, and her breasts have grown to full adult size slightly earlier than the average: they exhibit the perfect roundedness required, but still retain the firmness of extreme youth. This special combination has provided the kind of image of which centrefolds and men's-magazine fortunes have been made.

Once the visual signals of the female breast – and her other physical and mental charms – have attracted a male partner and sexual contact has begun, the tactile qualities of the breasts come into play. In pre-copulatory sequences there is often a great deal of oral and manual caressing of the breasts by the male. This excites the male even more than the female and it is possible that a special additional stimulus is operating here. It was mentioned earlier that the brown patches of skin around the nipples contain glands that secrete a fatty substance during lactation. This is claimed to be a soothing lubricant for the over-worked skin of the nipple region and there is no reason to doubt this. But the fact that the glands of the areolar region are, in origin, apocrine glands suggests that during sexual activity the nipple zone of the breasts of the female may actually transmit scent signals to the male nose. Apocrine glands are the ones responsible for the special sexual fragrance of the armpits and the genital regions, and although males are not consciously aware of the erotic odours these glands produce, their secretions do make a massive unconscious impact that aids sexual arousal. The areolar glands may well be part of this primeval odour-signalling system, which may explain why males exploring their partners' bodies spend so much time nosing around in the mammary zone.

As sexual arousal mounts, the female chest undergoes several marked changes. The nipples become erect, increasing in length by up to a centimetre. The breasts themselves become engorged with blood, increasing their overall size by up to 25 per cent. This turgidity has the effect of making their whole surface more sensitive and responsive to the body-to-body clasping of the mating pair.

With the approach of orgasm two further changes occur. The areolar patches become tumescent and swell so much that they start to mask the nipple, giving the false impression that the strongly aroused female is actually losing her nipple erection. There also appears a strange measles-like rash over the surface of the breasts and elsewhere on the chest. This 'sex flush' was observed to occur in 75 per cent of women who were the subjects of a detailed sex investigation. It is far less common in men but was seen in 25 per cent of those taking part in the same investigation. It is most likely to occur during the moments immediately before orgasm in both sexes. In females, however, it may sometimes appear quite a while before orgasm is reached, whereas in males it never appears until the last moment. Although it is not possible to develop this rash without experiencing intense sexual arousal, the converse is not true. Many individuals of both sexes never display the sex flush despite vigorous sex lives full of orgasmic experiences. Why people should differ in this way is not known. One important factor that favours the rash is a hot atmosphere. In cool conditions individuals who might otherwise display the sex flush do not do so. When it is very hot, on the other hand, the rash may extend far beyond the primary chest region, covering an area that ranges from the forehead to the thighs.

One of the facts that we all take for granted is that human females have only two breasts, but this is not always the case. About one in every 200 women has more than two. The condition is referred to as *polymastia*. There is nothing sinister in this and the additional breasts are usually non-functional. Sometimes they are little more than additional nipples, sometimes small breast-buds without nipples. Very occasionally a woman is found who does have more than two functional, milk-producing breasts. The most extraordinary case concerns a French woman presented to the French Academy of Medicine in 1886 by a learned professor. She had no fewer than five pairs of fully lactating breasts. A few months later, in one of the strangest medical contests ever, a rival academic was able to present a Polish woman who also had ten functional breasts.

These extra breasts are remnants of our very ancient ancestry: like most other mammal species, our remote forebears had several pairs with which they could feed a whole litter of offspring. When our litter size fell to one, or occasionally two, the number of nipples diminished in parallel.

Several famous women have possessed more than two breasts. Julia, the mother of the Roman emperor Alexander Severus, was many-breasted and was given the name Julia Mamaea as a result. More surprisingly, close examination reveals that the famous statue the Venus de Milo, in the Louvre, displays three breasts. This is usually overlooked because the third breast has no nipple and is little more than a small breast-bud. It is situated above the right breast, near the armpit. Henry VIII's unfortunate wife Ann Boleyn was also said to possess a third breast – a claim faithfully recorded in books on medical abnormalities. In this case, however, the alleged third breast may well have been a 'witchcraft' slur. Witches were once believed to have extra nipples with which they suckled their familiars, and women who were thought to be guilty of witchcraft were sometimes searched for telltale signs of their evil ways. Pious Christian witch-hunters would diligently search the most private crevices of a suspected witch for a hidden nipple. A wart or a large mole and sometimes even a slightly enlarged clitoris would be enough to have the wretched owner burned to death at the stake. Rumours of Ann Boleyn's third breast may have been deliberately spread at her death to suggest that she was evil and deserved to die.

The most famous multi-breasted figure in history is the Diana – or Artemis – of Ephesus. Her ample sculptured bosom displays several rows of breasts crowded together. Some versions of the statue show more than 20 breasts. Or do they? A closer look reveals that none of these breasts has a nipple or an areolar patch. They are all 'blind breasts'. Recently the cult of this ancient Anatolian mother goddess has been examined more carefully and an entirely new interpretation has emerged. To understate the case, the Diana's chest is a less friendly place than has long been supposed. It seems that the goddess's arch-priest had to be a eunuch: in order to serve her he had to castrate himself and bury his testicles near her altar. Inscriptions have been found which reveal that, after a time, bulls were substituted for priests at the castration ceremonies. Their huge testicles were removed and preserved in scented oils, then ceremonially hung on the chest of the sacred statue. The original statue was made of wood, but copies of it were made in stone, with the cluster of sacrificed bulls' testicles shown in place. It was a study of inaccurate stone copies that gave rise to the long-standing error that the Great Mother was many-breasted. The reason for covering the chest of the goddess with the testicles was that the millions of spermatozoa contained in them were thought to fertilize her. This was achieved in such a way that she could become a mother while remaining a virgin, a theme that was to reappear in connection with the birth of Christ.

A breast myth of an entirely different kind surrounds the ancient nation of female warriors known as the Amazons. It is doubtful whether they ever really existed but according to early writers they were a fearsome all-female community forever attacking neighbouring peoples with their bows and arrows. In order to release their arrows more efficiently they were said by some to burn off the right breast of all girls at puberty. Others had it that the offending breast was cut off. Unfortunately for these stories all the ancient works of art showing these ferocious females depict them with two sound breasts. If the Amazons existed at all it is more likely that

152

they wore a one-sided leather tunic that flattened the right breast during battle. The name Amazon means literally 'without breast' (a-mazos).

Curiously, in recent years Westerners have started to mutilate their breasts for erotic and decorative purposes. Cases are rare but sufficiently widespread to alarm sociologists, one of whom declared that the new fashion of 'erotic body-piercing of nipples, navel and labia, and the insertion of chains, jewellery etc' could easily forestall sensible legislation to outlaw the African custom of female circumcision. Modern nipple-piercing is essentially part of the bondage syndrome from the world of exotic sexual practices. Among tribal societies mutilation of the breast is extremely rare for the obvious reason that it interferes with breast-feeding – a serious drawback where bottle-feeding is not an available alternative.

Less harmful were the erotic nipple decorations found in earlier times. Three thousand years ago, in ancient Egypt, high-status women enjoyed covering their nipples lavishly with gold paint. In ancient Rome, 2,000 years ago, the preference was to decorate nipples with rouge, to spice up erotic encounters. The Empress Messalina, nymphomaniac wife of the long-suffering Claudius, was notorious for her red-painted nipples, as the satirist Juvenal was happy to point out:

Nightly she donned her cowl and with her maid, left him to
    play her shameless masquerade . . .
She bared her painted nipples and kept loose those thighs that
    birthed well-born Britannicus.

Among deliberate actions that transmit sexual breast signals are various forms of breast-cupping with the hands, chest protrusions and dance movements that shake or emphasize the shape of the breasts. All these draw attention to the sexual hemispheres of the female. The most extreme was the famous tassel dance of old-style burlesque shows in which the performers rotated both breasts in the same direction and then in opposite directions, with the tassels following suit.

The simplest form of sexual breast display is, of course, their full exposure in a context where they are expected to be covered up. This applies in urban societies throughout the world. 'Going topless' is a risqué action that always attracts a great deal of male attention. Sometimes the males in question have been uniformed police, as happened on the beaches in the South of France in the 1960s when young women decided it was time to venture out in topless bathing suits called the mono-bikinis, or monokinis, in order to gain a more extensive suntan. For a while there were unseemly struggles with embarrassed policemen grappling with near-naked women, but before long the authorities had lost the battle and topless bathing gradually became commonplace.

This first topless female bathing suit was introduced by controversial Austrian designer Rudi Gernreich in 1964. In the United States one of these suits was obtained by a nightclub dancer and used for her act, creating the first topless performance. Other clubs soon followed suit, but the following year religious opposition grew and the police raided the clubs and arrested the topless dancers, citing 'lewd conduct'. They were acquitted and immediately returned to work. In 1966 some New York restaurants introduced topless waitresses, but within a few days the Mayor of New York had them outlawed. In 1969 Ronald Reagan took a similar step in California. It was not until the 1970s that resistance to topless performances began to wane. Even then, limits were set and rules laid out as to how, when and where they could occur.

Curiously, something as natural and non-sexual as breast-feeding sometimes created a scandal in urban environments, when it was done in a public place. In 1975 three American women were arrested for breast-feeding their babies in a Miami park. Their offence was classified as 'indecent exposure'. Objections to such arrests mounted in the following years and today breast-feeding in public is legally permitted in much of North America.

In the 1980s a new approach to topless females in public was observed. Groups of young women, demanding full sexual equality, deliberately exposed their breasts in public places, insisting that they should be treated in exactly the same way as young men, who were allowed to take off their shirts without criticism. (In a similar vein, young men refused to wear collars and ties in upmarket restaurants because women did not have to do so.) This extreme form of sexual equality was not exactly what social reformers had in mind when they tried to stamp out gender abuses.

As the twentieth century drew to a close, naked breasts were increasingly exposed in newspapers and magazines, in the cinema and eventually on television. In lap dancing clubs they were literally waved under the noses of the male clients. Their overt visual impact is clearly still operating, but some of their tantalizing mystery has been lost.

It should be stressed that this more relaxed attitude to topless displays is largely confined to the Western world. Even in the twenty-first century, Westerners visiting foreign holiday locations have found themselves in difficulties when they overlooked this fact. As recently as 2003, a British teenager was sentenced to eight months' imprisonment, or a €2,800 fine, for exposing her breasts during a 'Eurovision Thong Contest' in a nightclub on the Greek island of Rhodes. She was accused of 'insulting local values and morals'. The taboo continues.

Before leaving the subject of female breast exposure, there is one extraordinary fact that demands mention. Uniquely, it concerns a law that was passed insisting that naked breasts *must* be shown in public – the complete opposite of all the other legal steps taken in this sphere. This law was passed in Venice, in the fifteenth century, and applied to prostitutes who sat in windows trying to attract customers. Homosexual practices were so popular at the time that some of the women had taken to cross-dressing in order to lure young men seeking male partners. This so outraged the authorities, who were trying to stamp out sodomy (which was punishable by death), that they introduced a ruling that, when working, the female prostitutes must always fully display their bare breasts to prove their gender. When they left their homes, there was one special bridge on which they stood and where they displayed their bodies naked from the waist up. The bridge became so famous for this that it was named the Ponte delle Tette – the Bridge of Breasts.

A brief word is necessary to clear up a misunderstanding about ancient figures that are shown squeezing their breasts with their hands. These have always been called Mother Goddess images and it has been supposed that they were emphasizing their breasts by cupping them erotically. We now know that this was not the case. These figures, usually found in ancient tombs, were mourner figures. In early times females performed a mourning ritual that involved both beating and squeezing their breasts. A side effect of this, if they were lactating at the time, was that the squeezing sometimes squirted milk from their nipples in long jets. It is possible that this action eventually became incorporated into certain rituals. Anthropologists have found, to their surprise, that in certain remote tribal societies, lactating females react in a similar way to sudden shock, grabbing at their breasts in panic and squirting milk several feet in front of them.

Finally, there is the inevitable question of how women can improve their breasts to make them transmit younger and/or sexier signals. For centuries, tight corseting has been used to push the breasts inwards and upwards, but although these garments improved the breast-shape of a woman, they also restricted her movements. When young women started to demand a more physically active role in society, there was an urgent need for clothing that permitted this. One of the first steps in this direction was taken at the start of the twentieth century, when the suffocating corset was divided into an upper and a lower portion – the brassiere and the girdle. Later in that century the girdle also disappeared, but the brassiere was here to stay. In 1935 it became known simply as 'the bra' and the combination of 'bra and panties' remains with us to this day as the favoured form of female underwear.

There is much debate among fashion historians as to who invented the brassiere. The New York socialite Mary Phelps Jacob (known professionally as Caresse Crosby) insisted that is was she who had invented it, having obtained a patent for it in 1914. The idea had come to her the previous year when she was dressing to go to a formal event and found that her corset, stiffened with whalebone and steel rods, was incompatible with the plunging neckline of her sheer evening gown. With a creative flourish, using two handkerchiefs and some pink ribbons, she put together what she claimed was the first bra.

In fact, she was merely reinventing it, because 'breast supports' had already appeared in France at the end of the nineteenth century and had been given the name of 'brassiere' as early as 1907. The French couturier Paul Poiret insisted that the honour of introducing the bra was really his, declaring that 'In the name of Liberty I proclaimed the fall of the corset and the adoption of the brassiere . . . I freed the bust.' And he was not the only one. The English couturier Lucile (Lady Duff-Gordon), who introduced the term 'chic' to the fashion world, counterclaimed that it was she who, in 1911, 'brought in the brassiere in opposition to the hideous corset'.

The simple truth is that they were all part of a general trend, as the twentieth century began, that saw the gradual freeing of the female body from its past restraints. And they gained encouragement from an unlikely source. During the First World War, the US War Industries Board became alarmed at the amount of metal being wasted to make corset-stiffeners. They began a campaign to stop women wearing corsets and in this way hastened the switch to bras. They later claimed to have saved 28,000 tons of metal by this means, 'enough to build two battleships'.

The newfangled brassiere had two quite distinct functions. It protected large breasts, preventing them from bouncing awkwardly during rapid body movements, and it also made them look firm and round, and therefore more sexy. When a few of the early feminists were reported to be burning their bras in the late 1960s, they were objecting to the second of these two functions.

Some feminist historians have claimed that bra-burning was little more than a publicity stunt by anti-feminists to belittle and ridicule the movement. This is surprising because, although the actual incineration of bras may have been exaggerated by the press, there was a period in the late 60s and early 70s when there was, indeed, a strong anti-bra movement that went hand-in-hand with a rebellion against elaborate make-up, vivid lipstick, tight shoes and other forms of direct female sexual display. The feeling at this time, when feminists were campaigning vigorously to ensure that women should be treated as social equals, was that men should accept female bodies as they are, without any embellishment, and because breast-boosting bras were part of this embellishment, they had to go. This phase did not last long, however, because the discomfort of a braless

157

condition was unacceptable to the majority of women and bra-burning was quickly forgotten.

In its erotic role, the bra has nearly always been designed to create a roughly hemispherical shape, but there was one curious period in the 1950s, when designers used it to create not a rounded but a pointed breast. This was achieved by introducing 'torpedo-shaped bras that are stiffened and built to sharp points, wholly defying nature and gravity'. These were augmented by the addition of pads called 'falsies'. The general result was an almost aggressively thrust-out chest that would soon give way to the softer roundness of the 1960s breasts and which would never reappear in ordinary daywear. We were to catch sight of them again only when they enjoyed a brief, conical outing in a 1994 performance by the singer Madonna, where pointed breasts re-surfaced with all the demure femininity of a pair of rocket warheads.

According to Hollywood legend, one of the most sophisticated bras ever designed was that created by billionaire Howard Hughes for Hollywood actress Jane Russell. For a particular film role, he wanted her to display super-erotic breasts without going topless. To achieve this he enlisted the services of a structural engineer who was a specialist in designing bridges. This man produced a set of plans from which it was possible to develop a specially cantilevered brassiere that lifted and at the same time separated her breasts. The result was so impressive that there were serious attempts to have the film banned as obscene. (That, at least, is the endlessly repeated story, but recently an elderly Jane Russell insisted that, in reality, she never wore the famous bra.)

Tight corsets and modern bras can help to emphasize the breasts, but once the clothing comes off, the game is up and something more drastic is needed. Enter the plastic surgeon. The introduction of implants to make even the completely naked breasts stand round and firm, started to take off in the 1960s. The first insertion of silicone gel implants was carried out by cosmetic surgeons in Texas in 1963. The operation became steadily more popular in the 1970s and 1980s and by the 1990s there was a boom in this procedure, with as many as 100,000 women a year going under the knife for the sake of a more prominent bust-line. It has been calculated that, by the year 2002, more than 1 million American women had had their breasts enhanced by surgery. This is a staggering figure for any kind of cosmetic surgery and reveals the deep-seated need many women have to display this primeval female signal.

Unfortunately, these surgically aided breasts never look (or feel) completely convincing. They are somehow too perfect and they do not move as freely or softly as they should do, as their owner shifts her body from one position to another. As a result of this, the twenty-first century has seen the beginning of a reversal in this trend. In 2001 no fewer than 4000 American women subjected themselves to another bout of surgery in order to have their silicone implants removed. This has alarmed some plastic surgeons, who have grown rich as the creators of super-busts, but it would seem to be a major and growing shift back to natural breasts, even if these are smaller than average.

The hope has been expressed that, in the post-feminist phase, men are beginning to choose their mates more on personality than on breast-shape, but sadly this is not always the case. Some women admit freely that they are having their breast implants removed simply because they have done their job. Initially, they helped these women to acquire a high-status mate, but once the brides were cosily settled down to married life they no longer required such powerful sexual signals and therefore discarded them.

Some women, rather naturally, have resented the fact that they had to undertake this type of surgery to please a prospective mate. One female lawyer summed up her surgical 'reversal' succinctly by saying that, following a divorce, 'the first things to go, apart from his smelly dog, was the bloody bosom . . . It felt like my IQ jumped by 20 points.'

15. THE WAIST

One of the key gender signals in identifying the adult human female is the hourglass shape of her torso. This silhouette is defined by a special feature of the female body – its slender waist. This narrowness of the waist is partly due to the broad features above and below it – the protruding breasts above and the wide, childbearing hips below. But even without these contrasts, the female waist is smaller than that of the adult male.

The usual way of expressing the 'waist indentation' of a body is to give it as a simple waist-to-hip ratio. The figure deemed attractive for an adult human female is 7:10; for a male, 9:10. The difference between these two figures is remarkably resistant to cultural changes. If people in one society find fuller figures appealing, or in another country like unusually slender forms, this does not affect the gender differences in waist-to-hip ratio. Men and women, whether chubby or spindly, still show a marked difference in their 'waist indentations'. This feature of the human body appears to have an ancient and very basic significance.

Today's women, largely free of waist constraints, have an average waist measurement of about 71 cm (28 inches). Young women specially selected for their slender beauty, such as models, beauty pageant contestants and magazine pin-ups, have an average waist measurement of about 61 cm (24 inches). Modern female athletes, selected for pseudo-masculine muscular strength rather than beauty, have thicker waists averaging 74 cm (29 inches).

Of course, to appreciate the 'contour value' of these figures they have to be related to the bust and hip measurements above and below them. It is the relationship between these three sizes that gives the all-important 'waist indentation' factor of the typical female silhouette.

A carefully selected beauty pageant contestant will have a perfectly balanced contour, with identical bust and hip measurements. Typically, a beauty queen will measure 91–61–91 cm (36–24–36 inches). A skinny fashion model, of the kind favoured by modern couturiers is more likely to measure 76–61–84 cm (30–24–33 inches). Such a model may have an exquisite face and make an elegant clothes peg, but she will lack the hourglass contour that appeals to the primeval eye of the sexual male.

The typical British woman has a slightly different problem, her average statistics being 94–71–99 cm (37–28–39 inches). Her hips, being 5 cm (2 inches) larger than her bust, give her what is called a 'two-inch drop'. This hippy (as distinct from hippie) condition is even greater in some other European countries. In Germany and Switzerland it is 6 cm (2.4 inches) and in Sweden and France it is 8 cm (3.2 inches).

This large-hipped condition is reversed in the pin-up model. The typical pin-up has measurements of 94–61–89 cm (37–24–35 inches). In place of a two-inch drop there is a two-inch rise. Her breasts are the same size as those of the typical European woman, but they appear bigger because her waist and hips are smaller. She will inevitably be described as being 'big bosomed' but this is only an illusion, created by her small waist and hips.

It could be argued that female 'vital statistics' such as these are outdated and irrelevant. The organizers of most beauty contests no longer dare mention them in post-feminist society. But the fact remains that they do continue to play a vital part in human relations. In a recent experiment a row of life-sized, cut-out female silhouettes of varying proportions was set up in a shopping mall and adult males who were passing by were asked to indicate which one they most liked. The vast majority selected the one with curvaceous, waisted, balanced proportions. The verdict of these randomly selected males clearly supports the view that a 'waist indentation' factor is still important. It is too deeply ingrained in the male psyche to be swept away by modern cultural attitudes.

As with other parts of the female body, the gender difference in waist size has led to artificial exaggerations. If a small waist is feminine, then a tiny waist must be super-feminine, and many a young woman has, in the past, been prepared to suffer to achieve this improved condition. The special appeal of an unusually small waist lies it its double message – its comparative smallness in contrast to the swelling breasts and hips that border it, and, in addition, its actual, absolute smallness.

The reason that absolute smallness has such appeal is simple and biological. After a young woman has given birth to her first offspring, there is always some degree of thickening in her waist region. Even if she manages, by strict dieting, to keep the rest of her body as slim as it was before she first became pregnant, her waist will never again be quite as trim as it was. This is because of the irreversible changes that take place in her abdominal region when she becomes a mother. It has been estimated that, following the birth of several children, the average female waist, if left unconstricted by tight garments, will have increased in size by 15–20 cm (6–8 inches). As a result of this, the tiny female waist has, for centuries, stood as a symbol of nubile virginity – of the woman who is ready for sex, but has not yet had it. This condition has such a huge, primeval attraction for the reproductive male of the species, that many women, even those who have left it far behind them, have yearned to recreate it, if only in a symbolic way.

To achieve this, waists have, for hundreds of years, been squashed and squeezed into strong belts, tight bindings or laced corsets and the use of these restricting garments has given rise to much heated debate. The arguments are far from simple. It is not a question of the puritanical versus the hedonistic, as with so many female fashions. Here, there are arguments both ways from both sides.

Those who have strongly opposed the cult of the tiny corseted waist include both the pious and the liberated. Back in the seventeenth century it was the pious who led the attack. They vigorously promoted the view that any attempt to improve on nature, with a woman's body, was an offence to God. Writing in 1654, John Bulwer raged against the 'dangerous fashions and desperate affectations about the waist'. He described a tight corset as a 'fashion pernicious beyond imagination' and issued dire warnings to young women who 'straight-laced themselves to attain a wand-like smallness of waist, never thinking themselves fine enough until they can span their waist [with their hands]'. If they ignored him they would, he thundered, 'soon purchase a stinking breath and . . . open a door to consumptions and a withering rottenness'.

This view was often repeated in the years that followed. The subtitle of a book on the perils of tight lacing, by the American Orson Fowler, published in 1846, referred to 'the evils inflicted on mind and body by compressing the organs of animal life, thereby retarding and enfeebling the vital functions'. In place of Bulwer's withering rottenness, Fowler promised the corseted woman insanity and depravity.

Other, less extreme critics also voiced their fears about medical complications brought on by the powerful squeezing of the tight corset-strings. Among the ailments listed were: headaches, fainting, hernia, liver damage, miscarriage and difficulties with both breathing and blood circulation. Some even went so far as to include skeletal deformities, cancer, kidney diseases, birth defects, epilepsy and sterility. One Victorian author listed no fewer than 97 medical conditions that he claimed were caused by wearing tight corsets.

All these health warnings were largely unnecessary because most young women who wore corsets were sensible enough not to go to extremes of tight lacing, or to wear their corsets continually over long periods of time. Clearly, fanatically tight lacing could restrict both respiration and circulation and cause headaches, fainting and gasping for breath. Prolonged tight lacing could also weaken back muscles, so that backache occurred when the corsets were finally removed. But limited use and moderate lacing could easily be employed to create an attractively small waist for special occasions, without causing undue medical damage, and this is how the majority of young corset-wearers behaved, despite horror stories to the contrary.

A completely different attack came from the liberated woman of modern times. For her, the idea of wearing any kind of restrictive costume was an insult to female freedom. To be restrained physically was not merely unathletic, it was also symbolic of a mental imprisonment by the male. The tight corset was an instrument of torture imposed on submissive women as part of masculine oppression.

If the modern, hedonistic female wanted to undulate her body provocatively on the dance floor, she could tolerate no kind of costume rigidity. If she wanted to engage in sexual equality during foreplay, she had to be as flexible and 'flesh-free' as her male partner. If she wanted a small waist, it would have to be gained by working-out in the gymnasium, by jogging and other vigorous health regimes, rather than by the passive solution of strapping on a tight corset. She would have to replace inactive clothing disciplines with active athletic disciplines in her quest for male admiration.

The intelligent feminist also wanted body freedom, but for a different reason. For her, the aim was to divert male attention away from her sexual body and focus it instead on the qualities of her brain. If she was going to impress a male partner, it was going to be with her intellectual ability and not her reproductive potential. So any attempt to exaggerate her feminine silhouette was taboo.

These, then, are the anti-corset voices, raised against any attempt to improve on the naturally occurring, curvaceous female contour. Ranged against them are the tight-lacers, again with several distinct viewpoints.

First, there is the opinion that tight lacing proudly displays self-discipline and symbolically stands for laudable self-restraint. It has, after all, given the English language the term 'straitlaced', meaning 'precise and puritanical'.

Second, there is the view that a tightly laced waist demonstrates respectability and a high moral tone because it helps to make the wearer inaccessible. She is virtually armoured against male attention. Her slim waist may excite masculine eyes, but her tight corseting with its complicated lacing makes her naked body far more remote.

In earlier days the corset was also important as an aid to aristocratic deportment. The tightly laced young woman is forced to adopt a stiffly erect, vertical posture of a kind that gives her an air of graceful aloofness. There is no slouching or relaxed sprawling for her. A special device that was employed to help to keep the torso vertical was a long, flattened piece of bone, called a busk, which was thrust down the front of the corset. (It was also said to be a useful weapon with which to defend the wearer against any male admirer who had lost control of his sexual urges and was attempting to rip open her lacings.) The opposite condition – the unlaced woman with her voluptuous freedom of movement – gave us the term 'loose woman'.

The tightly laced young woman also creates the impression (despite her concealed bust) of being slightly vulnerable – like a trapped animal. Her caged body restricts her ability to flee at high speed. This inevitably appeals to her male suitor, who can unconsciously enjoy the fantasy of how easy it will be to catch her, should he decide to pursue her.

For some men, this caged condition also lends a fetishistic appeal to tight corsets, endowing them with a strong hint of female bondage. In these cases, the sex appeal of corsets lies not only in the female silhouette that they create, but also in the unspoken knowledge that the admired female is suffering physical torture for her admirer. From here it is easy to see how the extremes of tight corseting have become an integral part of the sado-masochistic scene.

So, to sum up, both the puritan and the libertine can be found in the ranks of both the anti-corset and the pro-corset groups. The presence of a tight corset can make you either coldly straitlaced or tantalizingly sexy; and its absence can render you either a natural, liberated female or a loose woman.

Such is the interest in the smallness of the female waist that two popular myths have developed in modern times. The first is that in earlier days, as a result of tight corsets, minute waist measurements were frequent and widespread. There is a well-known claim that, in late Victorian times, towards the end of the nineteenth century, an attractive girl was one whose waist measurement in inches was the same as her age at her last birthday. A Spanish proverb demanded that a young woman should have a waist as slender as that of a greyhound. And there was an old saying that the ideal female waist is 'so slight that the sun can cast no shadow'.

It was widely believed that waist measurements as small as 38–41 cm (15–16 inches) were commonplace and that these were achieved by starting tight lacing early and continuing it ruthlessly throughout the years of puberty and young adulthood. Cartoons of the eighteenth and nineteenth centuries showed women being brutally laced until there was hardly any waist left at all. Recently, however, careful scholarship has refuted these ideas. The first blow was struck in 1949, when a close study was made of early garments and it was found that the smallest waist size that could be found in a huge costume collection measured 61 cm (24 inches). In 2001 a painstaking new study appeared that confirmed this. The smallest measurement found among eighteenth-century garments was 61 cm (24 inches). It is true that matters did become more extreme in the nineteenth century, thanks to the invention of metal eyelets that permitted tighter lacing, but even then, the smallest size was 46 cm (18 inches). Victorian waists, at the height of tight-lacing fashions, ranged widely from 46–76 cm (18–30 inches).

This does not mean that minute waists were non-existent, but that if they did occur they must have been isolated oddities rather than commonplace achievements. Even in modern times extreme examples have been recorded and the *Guinness Book of Records* mentions a twentieth-century English woman who managed to reduce her waist from a slender 56 cm (22 inches) in 1929, when she was twenty-four years old, to an astonishing 33 cm (13 inches) in 1939. She lived on for another 43 years after this, so clearly, in her case at least, the brutal squashing of her internal organs did her no immediate harm. But it is important to identify her as a freakish exception to the general rule, rather than an example of a major social trend. Women may want a small waist because of the primeval gender signal it transmits, but they will only go so far to achieve it. To go further is to become obsessive to a degree that damages the balance of life. A few of the tight-lacers went too far in earlier days, just as a few of their modern equivalents – the 'health diet fanatics' – have done today. But for the vast majority of the female population, these outlandish extremes were never attempted and suggestions to the contrary represent one of the great myths of fashion history.

The second myth is that, in pursuit of the perfect waist, Victorian women used to go to the horrific lengths of subjecting themselves to a dangerous rib-removal operation. It has been categorically stated in serious books on the history of fashion that, at the end of the nineteenth century, some young women were achieving the perfect hourglass figure by having their bottom ribs surgically removed. No details were ever given, but photographs were sometimes included to illustrate the amazingly tiny waist that had been obtained in this way. Many later authors (including myself in *The Naked Ape* and Germaine Greer in *The Female Eunuch*) accepted this and repeated the assertion, using it as an example of how far women were prepared to go in improving on nature.

It now seems that we were misled. A detailed investigation by Valerie Steele of New York's Fashion Institute came to the blunt conclusion that 'There is no evidence at all that this practice ever existed . . .' She points out that there is no mention of rib removal in any history of cosmetic surgery and that, at the end of the nineteenth century, it would have been an extremely hazardous operation. Medical procedures at the time simply were not advanced enough to consider taking such a risk. Looking again at the photographs that are supposed to show de-ribbed women, it seems likely that the images have been retouched to make the small waists appear even smaller.

Despite this, the urge to believe in cosmetic rib removal is so strong that a whole new 'urban legend' has sprung up around it. False rumours have been circulating for several years that certain world-famous Hollywood stars have recently undergone the operation. It is claimed that, now we have the advanced surgical technology at our disposal, it is something that can and has been done to enhance the figures of these famous women. The names of at least seven major stars have been mentioned as being among those who are rumoured to have sacrificed their lower ribs to their quest for beautiful figures.

The reality is that there is no evidence to suggest that any of these difficult surgical procedures have taken place, and most of the stars involved have simply ignored the claims as ludicrous. The rumours surrounding one, the singer/actress Cher, were, however, so persistent that she was obliged to issue a total denial, submit herself for medical examination and sue a famous French magazine for repeating the rumour.

Although it now seems clear that neither Victorian ladies nor modern celebrities have routinely subjected themselves to this extreme measure, the question remains as to whether the act of removing lower ribs for cosmetic purposes has ever taken place. It is hard to be certain, but there is some evidence to suggest that it may have been done in very rare instances. A description of surgical procedures offered to transsexual males who wish to appear more feminine includes the statement that 'Rib Removal is occasionally undertaken to obtain a more pronounced waistline.' But this is followed by the warning that 'It is widely regarded as inadvisable.' The names of several American cosmetic surgeons who are prepared to perform this operation have been published, along with a quoted fee of $4,500.

In Hamburg, a young woman claims to have reduced her waist measurement from 51 down to 36 cm (20 down to 14 inches) by means of a series of tight belts, corsets and a rib-removal operation. She reports that she was in hospital for three days after the operation and that it was successful, enabling her to appear on television in Germany, Australia and America to show off her extraordinary figure. Her claim may be true, but we can be certain that, even if it is, it represents an isolated, extreme case. Statements that 'rib jobs were relatively common in the fifties' and other similar comments remain unsubstantiated. Routine rib removal, it now seems, is no more than a myth born out of endlessly repeated gossip. The significance of its persistence is that it reflects, not a surgical truth, but the tenacity of a male fantasy. The image of a small female waist appears to be indelibly imprinted on the brain of the human male.

16. The Hips

The broad, childbearing hips of the human female constitute one of the key signals of the feminine silhouette. Regardless of whether the waist above them is narrow or not, wide hips spell out the primeval message that this woman is capable of producing offspring. Only when a human society is in a phase where it favours youthful playfulness over fecundity and reproduction does it abandon the appeal of wide hips for the boyishly slender look.

Because the female pelvic girdle is wider than that of the male, the wide hip is one of the major gender signals. To be precise, the average female hip width is 39 cm (15.3 inches) while that of the male is only 36 cm (14 inches). This basic biological gender difference has led to a great variety of exaggerations and modifications. Today most women are prepared to rely on natural unmagnified hip signals, but in the past they have often made themselves the slaves of super-hips, and the victims of super-hip technology. The lengths to which some fanatics went are hard to believe.

In the sixteenth century, European dress stores were busy selling huge cumbersome 'hip cushions' looking like padded motor-tyres and almost as big. These were tied on beneath vast skirts to double the natural pelvic width. They made any kind of vigorous or athletic movement impossible and created a costume so heavy and tiring that the ladies of the day were incapable of any energetic activities.

The eighteenth century saw the invention of 'hip improvers' or bell-like panniers. These steel-structured undergarments, designed to give the impression that their owners had childbearing hips, held the skirts out so wide that women had to pass through doors sideways.

Turning from shapes to movements and postures, it is not surprising to find that most hip actions have a strong female bias. Walking styles that involve conspicuous hip sway or wiggle are so strongly feminine that they are employed as caricature elements in comically erotic theatrical performances. Every major sex goddess has swayed a sexy hip from time to time. But only males impersonating females, or outrageously camp male homosexuals, would allow masculine hips to undulate in this way.

The sideways hip-jut is equally feminine, or effeminate. This is the postural equivalent of the walking hip-sway. It is a slightly contradictory pose, being both contrived and relaxed. The message it transmits is 'look what nice hips I've got', but the off-balance, asymmetrical posture of the body fails to signal a clear-cut mood.

Many dance routines incorporate vigorous hip movements, such as the side-jerk, the rapid hip-shake and the hip rotation, and again these are usually the province of the female rather than the male. In the famous Hawaiian hula dance, young girls, their pelvic regions emphasized by grass skirts, perform a variety of rhythmic movements of the hips, swaying them, jerking them and rotating them in time to the music. Two special dance movements are the *Ami* and the *Around-the-island*. The *Ami* is a Hip Roll. One hand is raised while the other rests on the hip. The hips are then moved in a circle, first clockwise and then counter-clockwise. The movement called *Around-the-island* is similar except that the body turns a quarter-circle with each hip roll, going 'around the island' in four counts.

Perhaps the most important hip gesture of all is the hands-on-hips action called the akimbo posture. It is usually said to indicate authority, defiance or an emphatic mood, but it is more complex than that. Essentially it is an anti-social gesture. It is the opposite of holding out the arms to invite an embrace. It is, indeed, quite difficult to embrace someone who stubbornly retains an akimbo stance. When the hands are placed firmly on the projecting hips, the elbows stick out sideways like arrowheads pointing away from the torso. It is as if they are saying, 'Keep away, keep back, or I will jab you.' It is performed unthinkingly and automatically

170

when the mood is right, with the performer hardly aware that she has done it. And it is global in distribution.

The akimbo action occurs whenever the person concerned is in a mood of rejection. This is why it is characteristic of defiance. The woman standing at the door of her house, arms akimbo, is silently saying, 'Keep away, no one shall dare to enter.' This is also why it carries a mood of authority. The person who holds authority, and wishes to display it, must be seen to be standing apart from others, not sharing space and posture with them. The akimbo gesture of a dominant member of a group tells the others to keep their places.

The akimbo gesture is also used by individuals who have just suffered a setback. They may not be in a dominant mood but they are certainly not seeking fellow-comfort. The sportswoman who has just lost a contest immediately adopts the hand-on-hips posture, usually with her head slightly lowered, reflecting her lowered spirits. Her akimbo message is 'Keep away from me, I am so angry I don't want anyone near me.'

If a woman wishes to dissociate herself from a group of people on, say, her left, then only the left arm comes up into the akimbo posture. If there is a group on the right with which she feels some affinity, then the arm on that side will stay down. These half-akimbos, often seen at parties or other social gatherings, quickly reveal the ties that certain individuals feel with the others present.

A strange feature of the akimbo posture is that, despite its world-wide use, it does not seem to have a name in other languages. It is given a descriptive phrase such as 'fists-on-haunches' or 'the pot with two handles', but there is no single-word equivalent. This reveals the extent to which the posture is taken for granted. It is one of those common human behaviour patterns which we see every day and to which we react subliminally without ever analysing the body signal we are receiving. If it were a more conscious gesture like a salute or a wave, there would be words for it in every language.

Finally, there is the interpersonal contact of the hip embrace. Young lovers often walk for long distances together, progressing snugly side-by-side, with their flanks touching and with their arms around each other. Their embracing hands come to rest on their partner's hips. It is as if they are attempting to indulge in a full embrace and walk along at the same time, with the hip embrace as the compromise solution. It is not an easy double action to perform and hampers forward movement, but on such occasions the mobility of the couple is less important than their display of intimacy – which is done both for themselves and for the others around them. It acts as a powerful 'excluder' to anyone accompanying them or watching them.

As a tie-sign, this kind of embracing has a more potent message than the equally common shoulder embrace. Two males may indulge in a shoulder embrace, either standing together or walking along. It is friendly, but there is nothing so intimate about it that it suggests a sexual liaison. To embrace the hips of another person, however, brings the embracing hand much nearer to the primary sexual zone and makes the action more sexually loaded. For this reason males offer such an embrace only to females, unless of course the intention is to display their homosexuality in public.

A sample of hip embraces was analysed for gender differences and it was found that in the majority of cases only one partner was actively embracing at any one time. The other was permitting the embrace but was not reciprocating. In 77 per cent of cases the embrace was by a male to a female; in 14 per cent it was by a female to a male; and in 9 per cent it was by a female to another female. (Parents embracing their small children were excluded from this sample.) As anticipated, there were no males embracing males, but it seems that the taboo on same-sex hip-embracing is less strict between females than between males. In this respect, however, it does not differ from many other public intimacies, such as greeting kisses.

The big difference in percentages between male-to-female and female-to-male embraces neatly sums up the whole attitude of adults to the hip region of the body. The males are obviously much more interested in embracing the female hips than vice versa. Viewed in social terms it is now clear that the hips are essentially female attributes. Because of the broad childbearing pelvis which makes them such a strong feature of the female form they are almost as loaded with femininity as the breasts. Only when the male starts to make his urgent pelvic thrusts do hips begin to acquire a more masculine flavour.

17. THE BELLY

The female belly has always been a taboo zone, not because it is in itself unusually erotic, but because it is so closely related to the primary sexual region located just beneath it. Clothing that exposes it attracts the gaze down towards the genital region. In the Western world, everyday clothing has therefore traditionally covered this area, but in recent years (since late 1998 to be precise) there has been a fashion for exposing it by wearing low-slung jeans combined with unusually shorts tops. This has brought the female belly out of hiding and made it into a new focus of male attention. In Japan this fashion even has a special name – *heso dashi*.

The reason for the choice of this newly exposed 'erogenous zone' is interesting and has to do with a major change in everyday female clothing. As mentioned elsewhere, in the Western world there has recently been a large-scale shift from skirt-wearing to trouser-wearing on the part of young adult females. Today, over 80 per cent of women are to be seen walking down city streets wearing jeans or some other kind of trousers. The result of this is that female legs have largely lost their role as areas of exposed flesh and some new zone was needed as a replacement. Low tops that expose shoulders and breast cleavage have been employed in the past, but that solution had become too familiar. Something novel was required and somewhere, someone had the bright idea of wearing a top that was too short to reach the trousers. Suddenly a new erogenous zone was born and the fashion spread rapidly. Legs might be frustratingly fully clad but, as a compensation, female navels were now available for male inspection.

(For a while, at least – until the fashion cycle moves on yet again.)

The theory behind this type of fashion change was first introduced by German costume analysts in the 1920s. They explained that, in modern female fashion, there exists a law of Shifting Erogenous Zones. This law states that young women will always want to display a particular part of their bodies, but that this display will keep moving from one region of the body to another. As one bit covers up, another becomes exposed. There are two reasons for this. The first is novelty – each new exposure is exciting because it has not been seen recently – familiarity has not yet bred contempt. The second is modesty – if more than one body zone is exposed at the same time, the impact is too vulgar. So, to keep the exposure fresh, but not too exaggerated, the erogenous zone keeps on shifting from one area of the body to another, as fashions change. Right now, at the start of the twenty-first century, the emphasis happens to be on the belly.

One special advantage of this is that the recent fashion for navel-piercing is no longer hidden. One of the problems of body-piercing below head level has always been that only one's most intimate companions become aware of it. But now, at last, decorative navel studs and rings have been able to come out into the open. As a result they have become increasingly popular and have spread from being an oddity employed by a specialist minority to a widespread fashion across a much broader field.

Pierced navels have an obvious decorative appeal, but it is surprising that sexually active young women should wish to wear jewellery in such a vulnerable position. Vigorous face-to-face sexual contact would seem to be a problem here, with a high risk of navel-tearing as body moves on body. Some writers have called in 'umbilical vandalism', but, despite this, by the start of the twenty-first century, it was reported that navel-piercing was second in popularity only to ear-piercing.

What of earlier attitudes to this part of the female anatomy? In Victorian times it became impolite even to use the word belly and a substitute had to be found. Because the belly region contains the stomach and because the stomach is positioned high up in it, well away from the 'unspeakable' genitals, the Victorians decreed that a belly-ache should become a stomach-ache. This anatomical inaccuracy became so entrenched that it has survived into modern times even though Victorian prudery has long since been put behind us. In Victorian nurseries, even the word 'stomach' was considered too anatomical and was modified to 'tummy'. Stomach-ache became tummy ache in the 1860s and that term too has stubbornly survived, reminding us that the Victorian heritage still lurks in the background of our supposedly liberated society.

While one class was shifting the belly politely up to the region of the stomach, another group was rudely pulling it down to the genitals. With equal and opposite inaccuracy they spoke of the belly as though it referred to the region below the pubic hair, rather than above it. An early slang expression for a mistress was a 'belly-piece' and the male's penis was called a 'belly-ruffian'. An 'itch in the belly' was sexual desire; 'belly-work' was copulation.

A third inaccuracy was to use the term belly as synonymous with womb. In the days when female criminals were executed for certain crimes there was a well-known strategy called the 'belly-plea' based on the ruling that a pregnant woman was spared capital punishment. In most prisons there were men called 'child-getters' whose less than arduous task was ensuring that inmates were qualified to 'plead their bellies'.

The correct Anglo-Saxon meaning of 'belly' is the lower front part of the body, below the chest and above the genitals, containing the stomach and intestines and, in the female, the uterus. In medical parlance: the abdomen.

This region of the body boasts few surface landmarks. Apart from the navel, or umbilicus (of which more later), there is the midline depression called the *linea alba*. In a typical adult this runs vertically from the navel up to the lower part of the chest. If a slender, athletic body is viewed in a suitable side-light, the *linea alba* shows up as a narrow but distinct indentation in the flesh, marking the place where the muscles of the left side of the body meet up with those of the right. In young muscular individuals the line can be detected below the navel as well as above it. However, anyone running to fat (at any age) will find it hard to detect either below or above the navel.

The belly of the female is more rounded in the lower part than that of the male. The female's belly is also proportionally longer than the male's, with a greater distance between the navel and the genitals. The typical female navel is also more deeply recessed than its male equivalent, assuming that both individuals are of similar average build. These differences can be summed up by saying that the human female has a larger and more curved abdomen than the male, a feature that is often exaggerated by artists.

As females grow older their bodies tend to become heavier and their bellies more generous. If they over-indulge in what used to be called 'belly-cheer' or 'belly-timber' – in other words, food – they soon become ruefully or proudly big-bellied. In earlier days of shortage, large bellies were often worn with pride and ostentation and tribal girls were fattened up for their bridegrooms. The new body puritanism, with its craving for eternal youth, has changed all that. Now, flat bellies are the order of the day for all ages.

This change of belly-fashion has had one unusual side effect. It has altered the shape of the female navel. On plump figures the navel is roughly circular in shape, but on slender ones it is more likely to be a vertical slit. A survey of works of art showing the more generously proportioned females of earlier days revealed that the vast majority (92 per cent) displayed circular navels. A similar survey of modern photographic models sees this figure fall to 54 per cent. So today's slender females are six times as likely to have vertical-slit navels as their more voluptuous predecessors.

There is more to this navel manoeuvre than mere loss of weight, however. All that the slimmer body does is to make the vertical navel a possibility. Whether it is displayed or not ultimately depends on the posture of the model. Even the skinniest female can present a circular navel if she slumps her body forward. So modern poses, either consciously or unconsciously, seem to be aiming towards a greater emphasis on the vertical-slit navel. The reason is not hard to discover. Because it looks like a body orifice, the navel has always played a minor role as a genital echo. Its indented shape in the centre of the belly makes it strongly reminiscent of the true orifices situated below it. The female's genital orifice is part of a vertical cleft, while her anal orifice is much more circular in shape. It follows that a shift towards a vertical navel display strengthens the specifically genital symbolism. In glamour photographs where the true genital cleft is concealed, the photographer and his model can contrive to offer a subliminal pseudo-orifice as a substitute for the real thing.

If this sounds rather fanciful, it is only necessary to look back at what happened to the navel in the more puritanical phases of the twentieth century. In early photographs it was studiously painted out, the pictures being retouched to give the ludicrous impression that the female belly was completely smooth. This was done because, it was said, the navel was far too suggestive. Suggestive of what, was never mentioned.

With early films there was also shock and horror at the exposure of this part of the anatomy by dancing girls. An official letter from the censor to the makers of *The Arabian Nights* stated: 'Passed for adult distribution provided that all dancing scenes showing the dancing-girl's navel are cut out.' A second wave of film censorship, in the 30s and 40s, returned to this suppression of the navel. The notorious Hollywood Code insisted that naked navels were to be outlawed. If they could not be covered in clothing, then they must be filled with jewels or some other kind of exotic adornment. What seemed to outrage the puritans about films in particular was that the dancers were able to *move* their navels – to make them gape and stretch as they undulated their half-clad bodies. This took the orifice symbolism too far and had to be ruthlessly stamped out to prevent sexual hysteria in the auditorium.

No sooner had the Western world relaxed its censorship of the cinema navel than another assault was mounted. This time it occurred in the true homeland of the belly dance – the Middle East. There, with new cultural and religious moods sweeping the Arab world, nightclub performers were instructed to cover their bellies when engaging in what had now come to be called 'traditional folklore dances'.

It is clear from these restrictions that the navel does have the power to transmit erotic signals, even if to most of us today it appears to be a comparatively innocuous detail of the human anatomy. Sex manuals have noted its appeal, stressing its fascination to young lovers who are exploring each other's bodies. The comments in such books reinforce its role as a genital echo. For instance: 'It . . . has a lot of cultivable sexual sensation; it fits the finger, glans or big toe, and merits careful attention when you kiss or touch.' (From *The Joy of Sex*.) A popular pose in illustrated sex manuals shows a man probing his partner's navel with his tongue – a pseudo-penis inserted into a pseudo-vagina.

For some, interest in the erotic possibilities of the female navel has reached fetishistic proportions. An organization calling itself the US Navel Observatory has devised a whole classification for this small detail of the female anatomy. Not for them the simple separation of navels into vertical and round. In a report entitled *Navel Architecture*, they recognize no fewer than nine navel shapes, as follows:

The Vertical Slit – a rare type; graceful, feminine and erotic.

The Navette Navel – strong vertical elongation, but wider in the mid-section. Named after a navette-cut gemstone.

The Triangular Navel – a common type, but considered beautiful; shaped like an inverted triangle with convex sides. Often with a deep furrow from the apex to the interior.

The Almond-shaped Navel – considered by the Japanese to be the zenith of umbilical beauty.

The Circle – a rare type today, the perfectly round navel.

The Oval – one of the most common shapes.

The Cat's Eye Navel – more horizontal than vertical in shape, giving it the appearance of an eye.

The Coffee-bean Navel – a shallow oval 'innie' with two flesh protrusions inside it. A combination of the 'innie' and the 'outie' navel.

The Pierced Navel – the modern mutilated navel.

Although this report is meant to be a light-hearted examination of female navels, the trouble taken to make it an accurate assessment reveals the level of sexual interest that the humble belly button can generate. Indeed, this is not the only navel classification that has been made. A German psychologist has compiled his own list of shapes, claiming bizarrely that you can 'understand yourself through your navel'. He recognizes: The Horizontal Navel, the Vertical Navel, the Protruding Navel, the Concave Navel, the Off-centre Navel and the Round Navel.

Outside the sexual sphere, the navel has caused something of a problem in religious circles. For those who believe in the literal truth of ancient religious texts there is the thorny problem of whether or not the first human beings possessed navels. If these beings were created by the deity, rather than born of woman, there were presumably no umbilical cords and therefore no navels. Early artists had the dilemma of deciding whether to include navels in their paintings of Adam and Eve in the Garden of Eden. Most of them opted to do so and no doubt invented their own reasons for the existence of these first navels, but their decision led to an even greater problem: since God created human beings in his own image, God too must have a navel. Naturally this gave rise to the further intriguing question: Who gave birth to God?

The Turks have found their own unusual solution to the problem of the first navel. They have an ancient legend which explains that, after Allah created the first human being, the Devil was so angry that he spat on the body of the new arrival. His spittle landed on the centre of the belly and Allah reacted quickly by snatching out the polluted spot, to prevent the contamination spreading. His considerate action left a small hole where the spittle had been and that hole was the first navel.

A totally different symbolism sees the navel as the centre of the universe and it is in this loftier role that it is contemplated by Buddhist ascetics. To 'contemplate one's navel' has often been misinterpreted as a self-centred, inward-looking form of meditation when in fact it is the exact opposite. It is an attempt to obliterate the individual by focusing on the whole universe via its central point.

Returning from the navel to the belly in general, there remains the question of how the famous female 'belly dance' – the *danse du ventre* – originated. It was mentioned earlier that it is now rather primly referred to as a 'traditional folklore dance', but for once it is a tradition whose beginnings have not been 'lost in the mists of time'. Modern puritans might prefer in this case that they *had* been lost.

There are three main movements in the belly dance: bumps, grinds and ripples. Bumps are forward jerks of the pelvis. Grinds are rotations of the pelvis. And ripples are muscular undulations of the belly region involving expert muscular control. The first two are easy and commonplace. The third is the province of only the most skilled female performers. All three are active sexual movements. They began in the harem, where the overlord was usually grossly fat, hopelessly unathletic and sexually bored. To stimulate him sexually, his young females would have to squat over his recumbent body, insert his penis and then wriggle their bodies enticingly to bring him to a climax. This wriggling became an expert activity, with special movements of the female pelvis and contractions of the abdominal muscles to massage the great lord's penis. As an act of copulation it has been described as 'fertile masturbation'.

As time passed, the female pelvic movements became developed into a visual display to titillate and excite the master of the harem before copulation itself was attempted. Freed from contact with his sluggish body, the harem women were able to exaggerate the actions and make them more rhythmic. With music added the whole display soon became stylized into what was called the 'muscle dance', and what we refer to as belly dancing.

Some sources have suggested an additional element. They claim that certain of the movements represent not copulation but birth. It is pointed out that in many cultures, before a pregnant woman was turned into a doctor's patient, she did not lie down to give birth, painfully pushing against gravity, but instead adopted a squatting position using gravity to help her deliver her child. She assisted the birth by moving her abdomen in a rolling motion, bearing down hard as she did so. It is this element of parturition that is said to have been incorporated in the belly dance as the centuries passed. It became not merely a dance of mimed copulation by a vigorous young female straddling an indolent, corpulent male, but a symbolic enactment of both conception and birth – the whole reproductive cycle in one performance.

Whether this modified interpretation of the belly dance is correct, or whether it is an attempt to sanitize a purely copulatory dance and bring it into line with other 'folkloric' activities, is hard to say. In any case, the purification process has gone much further in recent years. A belly dancing instruction manual published in the 1980s introduces its subject with the following words: 'In its new role as a healthy physical art form, the emphasis is on its keep fit qualities.' The harem dancing-girl has become a gym mistress.

Despite the fact that the belly dance is now being promoted as 'an excellent form of therapy for tension and depression', the names that have been assigned to its various movements still give a vivid glimpse of its more erotic origins. They include: the pivoting hip rotation, the travelling pelvic roll, the undulating pelvic tilt, the heel-hip thrust, the backbend shimmy, the hip skip and the camel rock. Clearly, all has not been lost.

In its non-sexual symbolism, the belly, like the navel, has several roles. Its most widespread association is with the earthier, animal side of human life. Because the belly is connected with our appetite for food, it becomes linked with all our animal appetites. There is a Greek proverb that states: 'The vilest of beasts is the belly.' And another pronouncement from ancient Greece cries out: 'May God look with hatred on the belly and its food; it is through them that chastity breaks down.' This unflattering Western symbolism is at complete odds with Oriental symbolism that sees the belly as the seat of life. In Japan the belly is regarded as the centre of the body.

Ordinary, everyday belly gestures are few and far between, and because of its proximity to the genital region, the belly figures rarely in interpersonal contacts. When someone touches someone else's belly, the two involved are usually members of same family, lovers, or very old friends. Parents may pat their children's bellies when they have eaten well; a proud husband may gently pat the protruding belly of his pregnant wife to show his pleasure at her condition;

and lovers may lie quietly together with the head of one resting on the recumbent belly of the other.

Apart from these actions and a rare punch in the belly from an enemy, there is only one other important person-to-person contact in this category and that is the belly-to-belly intimacy of frontal copulation. Strangely enough this posture is the subject of one of the oldest recorded jokes known to mankind. In one of the ancient Sumerian texts dating from the third millennium BC the writer records with sad humour: 'Brick on brick was this house built; belly on belly was it torn asunder.'

18. THE BACK

The female back has often been ignored, both by its owner and by onlookers. Other parts of the body – especially the head, breasts and legs – are given much more attention and also attract far more interest. And yet the female back has an undeniable beauty. Even at rest, it is naturally more arched than the male's back, and if this curve of the female spine is deliberately increased, helping to protrude the buttocks, this immediately adds a sexier outline to the profile of the body.

Seen from behind, the contours of the back are, of course, strikingly different in the human female and the human male, with the lower back being wider in the female and the upper back wider in the male. So there is a gender contrast both from the side and from behind.

Occasionally the female back has figured powerfully in the world of erotic imagery. As mentioned when looking at the nape of the neck, the Japanese are especially fond of this region in terms of sexual appeal. The kimono is cut away from the back of the neck to a precise degree, according to the status of the wearer. If she is a married woman, the alluring line of her spine is barely suggested, but if she is a Geisha, the kimono is cut so that it stands away from the back of the neck. When she kneels down in front of her male companion, he is able to glimpse the whole length of her back, tantalizingly revealed by the stiffness of her costume.

In the West, costume designers have, from time to time, placed erotic emphasis on the back. If dresses are being worn with high fronts, then the interest may be shifted around to the rear, with low-cut lines exposing most of the upper back. Hollywood started a craze for this in 1932, when actress Tallulah Bankhead appeared in a backless dress that was rapidly copied by the daringly fashionable.

More extreme versions of this design, revealing the whole of the female back, have appeared on rare occasions, when fashion designers have found a brave client who is prepared to shock at some major public event. One of the first examples of this was Ungaro's famous jumpsuit of 1967 in which the entire back was exposed, right down to the top of the buttock-cleft. This cleft then produced a 'cleavage-echo' relating the lower back to the female's upper-chest region. It also gave the wearer the possibility of displaying her sacral dimples and her 'Lozenge of Michaelis'.

These dimples are female back details that in the past have sometimes aroused male ardour to the level of passionate obsession. One writer, closing in on this part of the female body, described it as 'that silky, succulent, mouth-watering area, right where the little back dimples are . . .'.

The dimples are less evident on the slender female figures so popular today but, when the more voluptuous shapes were in vogue, they became favourite talking points among the more sophisticated libertines. The two dimples, which appear as small indentations on either side of the base of the spine, just above the buttock region, are present in both sexes, but are much more distinct on the female back because of the fatty deposits in that area. In males they are so poorly differentiated that they are only visible at all in 18–25 per cent of cases.

The classical world was fascinated by these female dimples and ancient poets sang their praises. Greek sculptors also paid loving attention to them. It is possible that the sexual appeal of female dimples on the cheeks of the face owes something to the presence of these other dimples close to the cheeks of the buttocks.

The Lozenge, or rhomboid, of Michaelis is a diamond-shaped zone situated between the sacral dimples, which was also a focus of erotic interest in earlier days. It is named after a German gynaecologist, Gustav Michaelis, who spent an undue amount of time studying it. It is sometimes surrounded and defined by four dimples, instead of two, there being one above and below the Lozenge, as well as the usual ones on either side.

Full exposure of the female back has not always proved successful, however. One critic, faced with ballerinas wearing backless ballet costumes, was moved to comment that 'their backs look numb and scared of exposure, like snails that have lost their shells'. Clearly the emaciated, muscular bodies of modern ballerinas are not best suited to a full display of the back. Without its curve-smoothing, underlying layer of fat it is in danger of becoming too wiry and 'stringy'. Sexy back displays, it would seem, are better left to those with fuller, more rounded figures.

Turning from its sex appeal to its biology, the female back is the hardest-working yet least-known region of a woman's body. Ever since our remote ancestors stood up on their hind legs, our back muscles have been forced to work overtime, and it is a rare individual who does not, at some time in her life, suffer from a nagging backache. For most women this is the only time they ever really pause to consider their backs as a separate part of their anatomy. At other moments it is a case of 'out of sight, out of mind', and there are few women who could identify themselves in a 'backs only' line-up.

If any woman did take the trouble to have a closer look at her long-suffering back she would find a brilliantly assembled set of muscles and bones with the twin functions of body support and spinal-cord protection. The cord itself, which is about 46 cm (18 inches) long and just over 1 cm (two-fifths of an inch) in diameter, certainly needs protection. If anything serious happens to it, it is time to buy a wheelchair. The back securely encases it, first in a triple-layered sheath, second in a shock-absorbing fluid, and third in a hard, blow-resistant casing we call the backbone. In reality, of course, there is no such thing as the backbone – there are 33 bones, in a long series. These, the vertebrae, are of 5 different types: The cervical or neck vertebrae have an amazing degree of mobility, permitting all the various head movements so vital to scanning the world and protecting the face. There are 7 of these bones. The 12 thoracic or chest vertebrae are much less mobile because their main job is acting as an anchor for the ribs. The 5 lumbar, or 'loin' vertebrae, the heaviest and stoutest of all, have the task of carrying the greater part of the body's weight. It is here that the dreaded backache is most likely to strike.

The sacral, or 'sacred' vertebrae are fused together beneath the lumbar region to form the curved sacrum. This consists of five vertebrae but they now act as one. It may seem curious that this triangular bone at the base of the spine should be thought of as sacred, but in occult circles it is designated as the most important bone in the body and plays a special role in rituals involving divination by the use of 'holy bones'. It was also considered to be the bone containing the body's immortal spirit. To most people, however, there is something strangely perverse about locating the 'soul' at the bottom of the back. Perhaps this was the point, because it is the sacral bone that is ceremonially kissed at the witches' sabbath.

The coccygeal vertebrae are the smallest and lowest of the bones of the back. They are also fused together, to form the coccyx – all that remains of our primate tail. The naming of this tiny, pointed bone is even stranger than in the case of the sacrum, for the word 'coccyx' means 'cuckoo'. You may well wonder what possible connection there might be between our remnant of a tail and a bird like a cuckoo. The answer lies in the bone's unusual shape, which early anatomists claimed was reminiscent of a cuckoo's bill. Some of our body parts have acquired their names in delightfully eccentric ways.

The muscle system of the back is extremely complex, but consists of three main units: the trapezius of the upper back, the dorsal muscles of the middle back, and the gluteus muscles of the lower back. Most cases of backache are caused by straining these muscles in some way. Omitting special medical problems, women suffer from back pain for one main reason – lack of exercise in a civilized, urban environment. If the back muscles grow weak from disuse, they can easily fall prey to misuse, and they are misused by poor posture, by sudden, unaccustomed violent action and by tension.

Poor posture arises from certain work styles, where the body is asked to hold a particular position for hours on end. It also becomes a problem during the increasingly lengthy leisure time of the Western world, where soft furniture has spread into every home. During the many hours spent watching television, talking or reading, the under-exercised urbanite snuggles into her soft chair or soft bed for comfort, like an infant seeking the security of her mother's body. Mentally this embracing, gentle furniture creates a sense of safety and calm, but physically it is putting heavy demands on the back muscles, which struggle valiantly to keep the spine – literally – in good shape. This is especially punishing when the slumped or sprawled or curled figure on the soft surface is overweight. Pregnant women have come to expect backache as an almost inevitable hazard of carrying the weight of a child, but unusually plump individuals, who carry an almost identical load in the same region, are often surprised when they start to suffer from similar symptoms.

Picking up heavy objects by bending forward and employing the back like a crane is another classic piece of misuse that often tests the human back beyond endurance. Although, for athletic women who are gymnasium-fit, there is little danger attached to this type of manoeuvre, those who live a less physically active existence are at some risk.

Mental tension is another way of testing-to-destruction the human back. Body tensions caused by mental anguish and anxiety can lead to prolonged strain on the back muscles. Eventually they begin to ache and this causes even greater anguish, which causes . . . and so on, until a doctor is needed. This process often takes place almost unnoticed at first and may be initiated by emotional problems that preoccupy the brain so much that side effects are ignored until it is too late. Another cause of backache, it has been claimed, is sexual frustration, and greatly increased sexual activity has sometimes been suggested as a cure.

In the world of symbolism, the back has little part to play, except as the housing of the spine. The spine itself was seen as a replica of the primal cosmic tree, reaching up to the heaven of the brain. Macedonians believed that, as a corpse rotted, its backbone turned into a snake. Other interpretations of the human spine saw it as a road, a ladder or a rod. In medieval times, the 'essence' of the spine was held to be unusually beneficial, and anyone who had more than a fair share of backbone was considered to be full of good luck. For this reason it was thought to bring good fortune to touch the hump of a hunchback. This belief still lives on in parts of the Mediterranean where small, plastic talismans can be purchased depicting a smiling hunchback. It also survives in the phrase 'I have a hunch' about something, meaning that I have a feeling I will somehow be lucky.

The back is not one of the more expressive parts of the female body. Even the expression 'to get my back up' is based not on a human posture but on the arching of the back of an angry cat. A woman can, however, bend, stiffen, arch, slump or wriggle her back as her moods change, and champion body-builders can even ripple their backs.

Bending the back forward, which for some elderly women becomes a chronic and permanent posture when walking, is an essential part of the subordinate acts of bowing, kneeling, kowtowing and prostrating. The vital element of all these actions is the lowering of the body to match the low status of the performer. In earlier days the movement had to be extreme enough to expose the whole of the lowered back to the dominant individual. This was, in fact, the only way in which a back could be shown without causing offence. To turn the back when standing upright was considered an unforgivable rudeness, because it was an active movement of rejection. For this reason it was necessary for subordinates to vacate the presence of a Great One by walking out backwards from the room or royal chamber. Even today there is a remnant of this formal procedure, observable at a crowded party when someone twists his head and says 'excuse my back' to a friend who, in the crush, has come face to face with it. And sharply turning your back on someone to whom you have just been introduced remains to this day a major insult.

If turning the back is rude because it deliberately ignores a companion, stiffening the back is threatening because it suggests a bodily preparation for violent action. For this reason, the military are specially trained in back-stiffening so that, even when they are relaxed and at ease, they appear a little more aggressive than the average citizen. Stiffening the back also has the effect of slightly increasing the overall body height, a change that aids a display of dominance. Slumping the back, which occurs with depression (and incidentally gives the condition its name) transmits signals of loss of dominance by lowering the body slightly – almost as though one were observing an incipient bow of subordination.

When a woman makes contact with her own back, there are several characteristic actions. The simplest is standing or walking with 'arms-behind-the-back'. This is done with the knuckles of one hand clasped in the palm of the other and is a popular posture of high-status individuals, especially royalty and political leaders on formal occasions, when they are inspecting special displays prepared for them. The posture is one of extreme dominance because it is the opposite of the nervous 'body-cross' in which the arms are connected in some way across the front of the body as a kind of safety barrier. The hands-behind-the-back posture says that the person is so confidently dominant that there is no need for even the slightest frontal protection. Schoolteachers also employ the same posture when walking in their school grounds, demonstrating their own dominance in that particular territory.

Other reasons for making back-contact include the secretive use of hidden gestures, as when a little girl puts her hand behind her back in order to cross her fingers when telling a lie.

There are also several ways in which others can make back-contact, the most common of which is the proverbial 'pat on the back'. This is an almost universal way of demonstrating comfort, friendliness, congratulation, or simple good humour. The reason this act is so widespread, and always has the same meaning, is that it is a miniature version of that most basic of all person-to-person contacts, the embrace. When she is very young a little girl enjoys the embrace of her mother's enfolding arms, spelling total security and love, and the feel of gentle hands pressing against her back becomes one of the primary body signals of caring and friendship. She still indulges in full-blown body-embracing and hugging when she is adult, if the situation is sufficiently intense and emotional, but in less passionate moments she switches to a minor version – the pat on the back – which reminds the body sufficiently of the major gesture. Even a brief, gentle pat given to someone in distress acts as a powerful comforting device, out of all proportion to the simplicity and brevity of the physical contact, because of the ancient echo it sounds from our infancy.

Another common form of contact is the back-guide, in which a companion is steered by lightly touching the back instead of the more usual forearm or elbow. The back-guide is slightly more intimate because it brings the two bodies closer together as they move forward. A related form of back-touching which does not involve actual steering is the light hand contact which 'lets you know I am here' when two people are standing together, facing in the same direction.

Because of its large, flat expanse, the back is one of the most popular areas for detailed decoration such as tattooing. Magnificent demonstrations of the tattooist's art can be seen pricked painfully into the backs of brave women all around the world. These include a popular joke-motif tattoo that shows a hunting scene with horses and hounds chasing the fox down the length of the tattooed back, with the fox's tail just about to disappear between the buttocks.

19. THE PUBIC HAIR

Throughout their childhood girls enjoy the simplicity of having virtually hairless bodies, with the obvious exception of the long hairs on top of their heads. Then, with the arrival of puberty, matters become more complex. As their ovaries start to enlarge and hormone production begins, visible changes appear, including the emergence of their first pubic hairs, just above the external genitals. Usually this happens between the ages of eleven and twelve, although in rare cases it may start as early as eight or be delayed until fourteen.

In typical cases, between the ages of twelve and thirteen, there is a darkening and thickening of the pubic hair. Then, between thirteen and fourteen, the quantity of pubic hair increases and starts to form a triangular shape. By the age of fifteen, the growth of this hair is usually more or less complete and it now resembles the adult pattern.

When this development occurs, some girls find it faintly unpleasant. The idea of having a hairy genital region strikes them as being either 'animalistic' or 'masculine'. As children, their bodies were clean and smooth, now they are suddenly 'dirty and furry'. They may have seen little of adult pubic hair, which may have been hidden from their view by prudish parents and film censors, and it may come as something of a shock to them. They will have noticed that men have hairy bodies and this may also make them uneasy.

These comments may seem exaggerated to those who grew up in cultures where there was a 'liberated' family atmosphere, but they remain true for large numbers of developing girls. The proof of this

192

emerged, unexpectedly, when a scientific study was being made of animal loves and animal hates. It was discovered that, among British children reaching puberty, spider hatred showed a huge increase in girls, but not in boys. By the age of fourteen, the precise time when pubic hair is showing its fastest growth, hatred of spiders leapt up dramatically to become twice as strong in girls as in boys.

At first sight, this had no obvious connection to pubic hair, but when the girls in question were asked to explain why they hated spiders so much, they nearly always replied that it was because they are 'nasty, hairy things'. Boys, who are expecting to develop body hair, like their fathers, are far less concerned about this. If they were asked why they disliked spiders, they were more likely to reply that 'some of them are poisonous'.

The hairiness of the spider is symbolic rather than real. What the fourteen-year-old girl is seeing, when a spider runs across the floor, is the movement of its long legs – legs that radiate from its soft, central body. It is these legs that are seen as 'hairs', and the whole spider is unconsciously viewed as some kind of 'mobile, hairy tuft'. The fact that the fear of this object doubles at the very moment when girls are coming to terms with the fact that there is a 'hairy tuft' growing between their legs, is clearly significant. So, for every girl who is proud to find herself sprouting body hair, there is another that is disturbed by it.

In different parts of the world, there is considerable variation in the type of female pubic hair, from short to long, from sparse to dense, and from straight and soft to wiry and curly. Also, in colour and texture, the pubic hair does not always match the head hair. Many dark-haired women have lighter pubic hair, often with a red tinge. Most women have wavy or curly pubic hair, even when their head hair is straight. The main exception to this is found in the Far East, where the straight black head-hair is matched by pubic hair that has been described as 'black, short, straight and not thick but rather sparse . . . forming a somewhat narrow triangle with apex upwards'.

One of the first questions pubertal girls are likely to ask about their pubic hair is 'Why do I have it?' 'What does it do?' There are three answers:

First and foremost, the display of pubic hair is a visual signal. In primeval, naked times, it will have acted as a signal indicating that a growing girl is now a sexual adult. Its full appearance at age fifteen coincides with the start of ovulation, and the biological ability to breed. To a prehistoric male, the absence of pubic hair in younger girls will also have been an important signal telling him that they are too young to breed. The presence of visible pubic hair will have helped to trigger his sexual response, while its absence will have inhibited it. (It is this inhibition that is so curiously – and so unnaturally – lacking in paedophiles.)

A second function of pubic hair is to act as a scent-trap. Skin glands in the genital region secrete special pheromones – natural scents that adult males unconsciously find sexually attractive – and their fragrance persists longer in the densely curly hair than on smooth, naked skin. There is, however, a disadvantage to this primeval kind of scent-signalling. In prehistoric times, when naked skin was exposed to the air, the natural fragrances remained fresh. But today, with tight clothing enclosing the pubic region, it is all too easy for a lack of hygiene to lead to a bacterial decay of the scent gland secretions. The result is an unattractive body odour. This is why modern, clothed humans have to wash more often than tribal unclothed ones if they are to keep their ancient scent signals operating successfully.

A third function of pubic hair is that it supposedly acts as a buffer between the skin surfaces of the adult male and female during vigorous sexual contact, protecting the female *mons pubis* from abrasion. This protective role is often mentioned, and there may be an element of truth in it, but the adult female in modern times who has had her pubic hair removed does not appear to suffer unusually from its absence, when her body is being subjected to a thrusting male pelvis.

In addition to these three serious suggestions, several other highly improbable functions have been proposed in the past. These include the idea that the pubic hair acts as a 'modesty concealment' of the genitals. Conversely, that it is a tantalizingly erotic veil that 'inflames the imagination'. Also, that it protects the genitals from cold and accident, that it absorbs perspiration dripping down the front of

194

the body, and that it 'facilitates the accumulation and mutual inter-change of electricity between the two individual opposite poles in copulation', whatever that may mean.

Perhaps the most delightfully odd observation of the use of pubic hair is that recorded by an early German anthropologist who visited the tribal people living on the Bismarck Archipelago in the South Pacific, where 'the women wiped their hands on their pubic hair whenever they were soiled or damp, as we are accustomed to use towels'.

As with many other parts of the female body, the pubic hair has not been left alone to enjoy its natural state. In both ancient and modern times there has been considerable interest in modifying it. This has involved dying it, shaping it, decorating it, or removing it, and, as always, there have been two strongly opposed views as to the acceptability of these modifications.

Those in favour of leaving pubic hair alone, in its natural state, include both the puritanical and the fun-loving. The prudes feel that to modify this part of the body in any way suggests an unhealthy obsession with sexual anatomy. Any shaping, trimming or colouring implies an interest in the visual display of body parts that should remain strictly private. Furthermore, they see the removal of pubic hair as the removal of something that helps to mask and obscure the vertical genital slit. In a hairless condition, this slit is fully exposed to view and heightens the gender display of the female body.

The early feminists saw pubic hair 'dressing' as pandering to males and rejected the idea, along with all other forms of make-up or cosmetic improvement.

Hedonists, in complete contrast, see the natural condition of the pubic hair as appealingly erotic because it presents the male with a primeval visual signal of female readiness to mate. In its role as a sexual scent-trap, it also offers the male the promise of a stronger retention of erotic fragrances from the female skin-glands.

The removal of pubic hair gives rise to two completely contra-dictory reactions. Puritanical support comes from the idea that pubic hair is potentially dirty and smelly, and that its removal is there-fore hygienic and cleansing. In addition, the idea of being completely smooth and 'having nothing between the legs', like a child's doll, is non-sexual and therefore non-erotic. This view has led, in the past, to many female statues having smooth pubic regions, with no hint of hair or any other genital features. It has also led to artists' models shaving off their pubic hair, ostensibly to clarify the details of their pelvic contours, but in reality to make them look more like sanitized classical statues.

There is one famous case where a naïve, intensely romantic Victorian scholar suffered terribly as a result of the artificial smoothness of classical statues. John Ruskin, Britain's first Professor of Art, was 28 when he started courting his future wife, knowing little of sexual matters. He married her the following year and she was stunned when she discovered that he was incapable of making love to her. After years of evasive action, he finally admitted that he found her pubic hair repulsive. As an ardent student of classical marble sculpture, he knew the naked female form intimately, and enjoyed it aesthetically, but he had never set eyes on female pubic hair and apparently had no knowledge that it even existed. (Classical marble statues of males show curly pubic hair, statues of females do not.) His horror on finding that his beloved partner displayed a hairy tuft between her legs was so intense that he was never able to consummate the marriage and his wife was eventually forced to have it annulled, despite the embarrassment of having to prove, through medical examination, that she was still a virgin.

If some puritanical males have shown a marked preference for a hygienically hairless female crotch, it is perhaps surprising to discover that many licentious males have shown remarkably similar interests. Just as the fully haired crotch has appealed to both the pure and the impure, so has the naked one.

The sexual appeal of pubic hair removal has three sources. The first is that removal of the pubic hair lays bare the vertical genital slit. In classical statues this detail is tastefully omitted; in traditional artists' models the slit is often obscured by the pose they adopt, or is again omitted by the artist in the finished work. In real life, however, this intimate detail ('cleft by God's golden axe') is fully exposed and transmits an even more powerful visual signal than the tuft of hair itself, to a watching male.

The second is that the hairless condition transmits a signal of virginal innocence. It is the body image of a girl too young to have sex and therefore, symbolically, too young to have *had* sex. Comments by males who have responded favourably to seeing a depilated female pubis, include the following telltale remarks: 'it is baby smooth', 'it is the schoolgirl fantasy thing', 'it's the Lolita look'. Critics have snapped back that it is 'a step in the direction of kiddie porn', but they overlook the fact that the men who are aroused by the hairless pubic display are aware that the rest of their female partner's body is adult. The fact that they like a symbolically 'virginal' feature does not mean that they would respond sexually to a genuinely pre-pubertal girl. One woman, defending her hairless pubis, remarked that 'any woman who suggests that a man's liking for a shorn vulva means he is a closet paedophile, risks having the same logic thrown back at her, unless all her lovers had full beards'. If nobody has accused women who prefer 'little boy' clean-shaven faces on their male lovers of having paedophile tendencies, then why should a clean-shaven female pubis be viewed in this way?

Apart from its innocent quality, several other advantages have also been mentioned in connection with the removal of female pubic hair. The genital region becomes much more sensitive to tactile stimulation. In particular, the pleasures of oral sex are greatly increased for both partners. Some women have reported that even the simple act of taking a walk feels more erotic: 'Just walking down the street is fun because you glide', it 'puts an extra swagger in your step'. Others see the appeal as being that of 'having a sexy secret, known only to you and your partner'.

To sum up, the contradictory attitudes towards female pubic hair include seeing full hair *puritanically* as natural and modest, or *licentiously* as erotically adult and fragrantly sexy; also, seeing removal of pubic hair *puritanically* as hygienic and cleansing, or *licentiously* as genitally exposed and sensitized. As with so many aspects of the adult female body, there are highly conflicting viewpoints.

Turning to the history of the removal of female pubic hair, it is clear that it is far from being a passing whim of modern fashion. Records show that it can be traced back as far as the ancient Egyptians. Egyptian women were fastidious about their body hair, removing all traces of it. This was done by a 'waxing' process, using a sticky cream made out of honey and oil.

It has also been claimed that King Solomon disliked female pubic hair. When the Queen of Sheba visited him, in the tenth century BC, he is reputed to have asked her to depilate herself before they made love, requiring that she made herself open to him by removing 'nature's veil'.

A little later, in ancient Greece, it is recorded that men preferred their women to 'remove the hairs from their privy parts'. This was because 'the strong growth of hairs of southern women would otherwise prevent their private parts from being seen . . .' As a result, in ancient Greece female depilation was the rule. It was achieved by one of three techniques. The first was hair-by-hair plucking; the second was singeing with a burning lamp; and the third by singeing with hot ashes.

The practice of female pubic hair removal was also popular in ancient Rome, but their techniques were slightly different. Like the Greeks, they did employ plucking, using special tweezers called *volsella*. Unlike the Greeks, however, they replaced the risky singeing techniques with the less hazardous application of depilatory creams. A form of waxing was also practised, using pitch or resin. Among fashionable Romans, young girls would start to employ one of these methods as soon as their pubic hairs began to grow.

When the Crusaders were in the Holy Land they discovered that Arab women depilated their pubic region. Impressed by what they experienced there, they brought the fashion back with them to Europe where some aristocrats adopted it during the Middle Ages. It flourished for a while but then died out.

Later, in the sixteenth century, it is known that Turkish women were so keen to denude their pubic region that special rooms were set aside for this purpose in the public baths. It was thought to be sinful for them to allow their pubic hair to grow naturally.

By Victorian times, in Europe, the removal of pubic hair was unheard of, except possibly among some of the 'ladies of the night'. It did not re-surface as a popular female fashion until the sexually liberated days of the 1960s. Then, suddenly, anything was possible and certain leading figures rebelled against customs considered to be prudish or traditional. One of the most famous rebels was designer Mary Quant, who set out to shock by announcing publicly that she had had her pubic hair trimmed into a heart shape by her husband. Others soon followed her lead.

During the 1970s the rise of the feminist movement saw a return to the natural look and pubic topiary once again fell out of favour. By the end of the twentieth century, however, it was back in a big way, with a whole variety of different styles. This new trend began because of a change in swimsuit fashions. The lower line of beach-wear was cut higher and higher (to make the legs look longer) and this resulted in a few stray pubic hairs emerging on either side of the now very narrow strip of cloth between the legs. These hairs looked ugly and were quickly trimmed away. This 'concealment' trimming set in motion an increasingly drastic reduction in pubic hair. More and more extreme styles appeared until, eventually, complete removal of every single hair became the ultimate quest of the fashionable. By the early part of the twenty-first century, extreme pubic depilation had become the latest female fashion fad, a defiantly modern trend which, paradoxically, has returned us once more the popular pubic styles of the ancient civilizations.

A whole new terminology has sprung up around this denuding craze, with each beauty salon having its own set of names for the different degrees of pubic nudity. These names are not always employed consistently, from salon to salon, but a rough guide is as follows:

The Bikini Line: This is the least extreme form. All pubic hair covered by the bikini is left in place. Only straggling hairs on either side are removed, so that none are visible when a bikini with high-cut sides is being worn.

The Full Bikini: Only a small amount of hair is left, on the Mount of Venus (the *mons pubis*).

The European: All pubic hair is removed 'except for a small patch on the mound'.

The Triangle: All pubic hair is removed except for a sharply trimmed triangle with the central, lower point aimed at the top of the genitals. It has been described as 'an arrowhead pointing the way to pleasure'.

The Moustache: Everything is removed except for a wide, rectangular patch just above the hood at the top of the genital slit. This is sometimes called 'The Hitler's Moustache', sometimes 'Chaplin's Moustache'.

The Heart: The main pubic tuft is shaped into a heart symbol and may be dyed pink. This is a popular style for St. Valentine's Day, presented as an erotic surprise to a sexual partner.

The Landing Strip: The central hair is trimmed into a narrow vertical strip and all other pubic hair is removed. This has become popular with models who must wear garments of an extreme narrowness in the crotch region.

The Playboy Strip: Everything is removed except for a long, narrow rectangle of hair, 4 cm (1½ inches) wide. This precise measurement may seem odd, but it has a legal history. In the American state of Georgia, exotic dancers were required to leave a strip of pubic hair that is 'two fingers' width' (which is equal to about 1½ inches or 4 centimetres) if they performed naked. This was considered, by Atlanta's lawmakers, to provide a sufficiently modest cover for the genital slit. A one-finger strip was considered to be obscene and was forbidden by law. Local lawmen were forced to undertake the arduous nightly task of checking for too-narrow strips of pubic hair and sending any girl home who disobeyed this strict rule. After a while the novelty of this strange duty wore off and the rule was relaxed. Confusingly, some salons refer to *The Playboy* as meaning 'everything off'.

The Brazilian: This is the most famous of the new styles but there is some confusion about its exact form. To some it is the same as the Landing Strip, to others it is a more extreme form of the Landing Strip, leaving only a 'vertical stripe of hair'. To still others it signifies the removal of all pubic hair. It began on the Copacabana Beach in Rio, where the highest cuts in bikini-lines (little more than thongs) first appeared. Then, a family of seven Brazilian sisters (known as the J. Sisters) moved to New York and opened a Manhattan beauty salon, where they started offering pubic depilation to their clients. Movie stars and top models began to visit them and soon their salon became the Mecca for pubic waxing. It was because of the J. Sisters' growing fame that the name of their style became known as the Brazilian. When other salons copied them they did not always follow the exact degree of hair removal, hence the confusion. But the J. Sisters have made it quite clear what they do, describing it succinctly as 'Everything off except a tiny strip'.

The Sphynx: This is unambiguously the 'everything off' style, leaving a completely hairless pubic region. The name is derived from that of a naked breed of cat from Canada. The smooth-skinned, hairless Sphynx Cat was a genetic oddity discovered in Toronto in 1966. Some salons refer to 'the Sphynx' as 'the Hollywood'.

These are the most popular pubic styles at the beginning of the twenty-first century. In addition there are:

Speciality styles: Some fashionable pubic-hair stylists offer exotic variants, such as a corporate logo; bull's eyes; stars-and-stripes; a target; or a family crest. One flamboyant stylist has introduced what could be called pubic kitsch, with titles like: honeymoon surprise, alpine jet stream, mod squad, queen of diamonds, cha-cha and flower blossoming. Another offers the exclamation mark, three-tipped leaf, crown, star, Mohawk, purple lightning bolt, or partner's initials.

To obtain these styles, tweezer-plucking, shaving, dyeing, depilatory creaming, electrolysis, sugaring and waxing have been employed. The most fashionable technique today is waxing, after which the pubic region remains hairless for several weeks until new hairs have had time to grow.

A complete contrast to pubic hair reductions is the curious custom of merkin-wearing. A merkin is a pubic wig made of human hair, nylon or yak belly hair. It is held in place either by an inconspicuous

G-string, or by gluing it on to the real pubic hair beneath.

Merkins have a long history, stretching back hundreds of years and are still offered for sale today. Their original function, centuries ago, was to mask the ravages of syphilis or other venereal diseases that disfigured the external genitals. Later they were used by prostitutes for customers who found a generously bushy pubic patch sexually appealing. More recently, in the world of film-making, they have been used as a 'modesty mask' by actresses required to appear naked in sex scenes. They have also been used as a temporary 'stick-on' alternative to pubic hair styling. The merkin can be given an exotic shape and colour for a special occasion, and then removed afterwards – a more convenient solution for some women than going to the lengths of genuine pubic topiary.

Some merkins have been decorated with jewels, flowers, or coloured ribbons and all three of these types of pubic embellishment have been known for centuries – both on merkins and occasionally even on real pubic hair. There are records that prove they were popular as long ago as the sixteenth century. Knowledge of their use was obtained in an unusual manner. The murdered body of a French marchioness was left in a public street with its genitals deliberately exposed. There, for everyone to see, was her patch of pubic hair 'adorned with crimp ribbons of different colours'. It seems that, when the King of France insisted that the ladies of the royal court should restrict the splendour of their costumes, they obeyed their monarch but then compensated for this imposition by taking their fashionable fripperies underground. Outwardly, publicly, they were obedient to the King's wishes, but secretly they continued to indulge in their ornamental excesses, vying with one another to create the most glamorous of pubic tufts – some tied with ribbons, some bedecked with flowers and some dotted with precious jewels. Those adorned with jewels sometimes rendered the pubic region the most valuable part of the female body and this led to a popular euphemism in which female genitals were referred to as 'a woman's treasure chest', or simply her 'treasure'.

20. The Genitals

With this part of the female body we have come to the major taboo zone. As a source of great sexual pleasure the genitals should be celebrated and yet they are rarely mentioned in polite society. (The brilliant play *The Vagina Monologues* is a unique exception to this rule.) Why should this be? Why should people be so reluctant to talk about this important region of the female anatomy? To find the answer it is necessary to turn the clock back to primeval days.

When our early ancestors first took to walking about on their hind legs they found themselves unavoidably offering a full-frontal display whenever they approached their companions. Previously it had been normal to advance on four legs, with the genitals completely concealed and well protected. To display the genitals then required a special posture. Now they were on show every time one human animal turned towards another. This meant that it was impossible for an adult to approach another without making a sexual statement. As a way of damping down these signals both males and females eventually took to wearing some kind of covering over the genital region: the loincloth was born.

The loincloth had three advantages. It not only reduced the strength of the sexual display of its owner when he or she was in a non-sexual public context, but it also intensified the sexuality of the private moments when it was removed. Thirdly, it helped to protect the delicate genital region from the harsher surfaces of the natural environment.

Today whenever people shed their clothes because of the heat it is always the modern equivalent of the loincloth that is the last to go. Unless we are dedicated nudists, we reserve our genital displays almost exclusively for our sex partners. Only with tiny children in a clearly pre-sexual phase do we relax this rule.

In most countries we do not rely on custom alone to keep adult genitals covered but also impose formal controls: it is against the law to 'exhibit' one's genitals in public. Generations of pious church-goers have responded to calls from the pulpit – 'Nudism is . . . as shameless as the Devil himself . . . the zenith of human rebellion against God.'

What precisely are we at such pains to conceal? In the case of the adult female there is relatively little to see. Beneath the pubic hair, and partly obscured by it, is a small vertical slit created by the pair of outer labia – fleshy folds that protect the more delicate inner labia which flank the vaginal opening. At the top of the slit is a small hood of flesh that partially covers the clitoris, a small and highly sensitive button of flesh just above the urinary opening, the urethra. And that is it. Compared with the male's sexual equipment, it can only be described as visually modest. Yet the attention it attracts is enormous and the lengths to which people have gone to conceal it from view have been extravagant, to say the least.

The reason for the excitement this body zone generates does not lie in its visual qualities, but its tactile ones. No other part of the female body is so sensitive to touch during sexual encounters, whether from the male fingers, lips, tongue or penis. The special design of the male's penis is significant in this respect. Compared with the penises of other primates, the human organ is most unusual. It lacks an *os penis* – the small bone that clicks into position to provide monkeys and apes with a rapid erection. Instead, it relies on vasocongestion to achieve erection. This is a system that, when sexual arousal occurs, allows blood to enter the penis more rapidly than it can leave. This not only stiffens the penis but it also greatly increases its length and, especially, its width. The result is that, when it is inserted into the female's vagina, it puts pressure on the vaginal walls and the labia. This pressure creates a strong erotic response in the female, permitting her to share the male's mounting sexual excitement as copulation proceeds.

This may seem an obvious and inevitable mechanism of mating, but it differs markedly from what occurs in other primates. Female monkeys receive a few rapid thrusts from the thin bony spikes of their males and in a trice copulation is over. In baboons, for example, a typical mating act takes as little as 8 seconds, with ejaculation occurring, on average, after only 6 pelvic thrusts. Even an abnormally long mating act would take no more than 15–20 seconds. As a result, female monkeys do not enjoy the slowly increasing sexual arousal and the explosive orgasms experienced by the human female. The thick human penis causes powerful contact sensations as it moves against the surfaces of the female genitals, during the often-lengthy pelvic thrusting of our species. The female orifice, surrounded by its highly sensitive folds of skin, is subjected to repeated, rhythmic massage by the tightly fitting penis. As female sexual arousal gradually grows, both the outer labia and the inner labia become engorged with blood, swell to twice their normal size and develop greatly increased sensitivity to touch. After prolonged stimulation the female eventually experiences an orgasmic climax that is physiologically very similar to that of the male. This means that both partners experience a massive reward for their sexual labours and the encounter, unlike that of monkeys, may lead to a tight emotional bonding between the pair. The fact that the human female (unlike female monkeys) gives no clear signal to the male when she is ovulating also means that the majority of copulations are not procreative but instead serve to further tighten the emotional bond. When human beings make love they *literally* make love.

Collectively, the external genitals are known as the vulva. It is worth taking a closer look at each part of this region, separately:

The Mount of Venus. Also known by its Latin name of *mons veneris,* and sometimes called the *mons pubis,* this is a sturdy pad of fatty tissue, covered in pubic hair, that acts as a buffer for the pubic bone. It is situated just above the labia, and its role is to protect the pubic bone from the impact of the male's body during the more vigorous moments of the sexual encounter. It also plays a role in sexual arousal because it is well supplied with nerve endings. Any accidental or deliberate massaging of this region will have an erotic impact and some women have claimed to be able to reach orgasm simply from *mons* massage. It is more sensitive to massage when its covering of pubic hair has been removed, which may partly account for the popularity of pubic depilation.

The Mount of Venus does not appear until puberty, when the sudden increase in levels of oestrogen triggers its formation. However, fashionably emaciated women will find it hard to develop this fatty tissue and, as a result, their vulvas will appear to be located further forward than usual.

The Outer Labia. Also known as the *labia majora,* meaning 'big lips', the fleshy outer labia normally cover up the inner labia, unless the legs are held wide apart. When they are close together they create a vertical slit between them that is referred to as the genital or pudendal cleft. A thin scattering of pubic hairs sprouts from their surfaces, and they are well endowed with scent glands. The skin is similar to that on the rest of the female body, although it may be slightly darker in colour.

During intense sexual arousal the outer labia may become redder. The male equivalent is the scrotum. There is some variation in the size of the outer labia, from woman to woman, there being more fatty tissue in some individuals, making their labia more rounded and prominent.

The Inner Labia. They are also known as the *labia minora*, meaning 'little lips', or the *nymphae*. Positioned inside the fleshy outer labia, these smaller, flatter (fat-free) lips take the form of a pair of sensitive flaps covered in a completely hairless, highly sensitive, mucous membrane that is kept moist by fluid seeping through from the blood vessels beneath the surface. During pelvic thrusting, these inner labia receive such prolonged tactile stimulation from the erect penis that they become swollen and suffused with blood, making them appear red in colour. (Absence of this redness in a seemingly orgasmic female is usually a sign of deception.)

There is considerable variety in the shape of these inner lips, some being small and smooth, while others are folded, ruffled, winged or lumpy. Among the Bushmen (San) people of southern Africa, the *labia minora* are sometimes greatly elongated and hang down between the legs 'like two flesh-finger pendants'. According to some reports, they can measure up to 11 cm (4.5 inches), and can be tucked into the vagina. One authority insists that they may even extend to 20 cm (7.9 inches) in length, and there is a hard-to-credit account dating from the 1860s that one mother 'was able to fold back her own *nymphae* so that they met behind over her buttocks. They have been referred to as the 'Hottentot Apron' or the 'Tablier Egyptien', and there has been a great deal of debate as to whether their greater length is a racial characteristic, or whether it is the result of a cultural custom of artificially stretching the labia.

Labia-stretching has re-surfaced as a modern Western practice in recent years and there are now clubs that teach it as a practice that increases sexual pleasure. There is no general agreement on this, however, some critics insisting that larger labia lead to painful chafing and tangling in tight clothing. Also, larger labia are considered ugly by some writers, who insist that 'Perfectly developed women will always have symmetrical *labia minora* which do not protrude past the outer lips; they will also have *labia minora* that are smooth and without excessive folds or crevices and wrinkles.'

Cosmetic surgeons would agree with this last view, since most requests for genital enhancement are concerned with reducing the size of the inner labia or, where one labium has become bigger than the other, having labial symmetry restored. Labioplasty, as this is called, has become the most commonly requested form of what is now known as 'intimate surgery'.

The Vagina. The vaginal passage is a tube of flesh about 8–10 cm (3–4 inches) in length when not in a state of sexual arousal. When at rest, its front and back walls touch one another. With sexual excitement it expands to a length of 10–15 cm (4–6 inches). During the adult phase, between puberty and the menopause, the lining of the vagina is slightly corrugated. Before and after this phase it is smooth.

In virgins, at the outer end of the vagina a thin membrane of skin, like a collar, partially closes the entrance. The presence of this hymen, or maidenhead, has been of extreme importance historically, in cases where bridegrooms have demanded untouched brides.

Traditionally, the skin is torn by the first insertion of the male's penis on the wedding night, and there is bleeding. In some cultures it became a significant ritual to display the bloodstained sheets of the marriage-bed as visible proof of the bride's virginity. Experienced women, faking honeymoon chastity, were known to insert a small sponge soaked in pigeon's blood into their vaginas, or to conceal a phial of animal blood under their pillow to spill on to the sheets at a convenient moment.

In modern times, with young females indulging in many kinds of vigorous sports, not to mention inserting tampons and masturbating in various ways, many hymens are torn before any penetrative sex has taken place. As a result, only about 50 per cent of modern women show the traditional bleeding when they first engage in vaginal intercourse. It has therefore been pointed out that, in today's society, 'virginity is a spiritual attribute and not a physical one'.

In evolutionary terms, the existence of the hymen is puzzling. Its effect is to render the first sexual act both painful and difficult. What survival value can this have? The only possible explanation seems to be that it was an evolutionary step that put a slight brake on early sexual encounters. Deflowering a girl became a greater threshold for a boy to pass over, and the first sexual act between a pair of young lovers became a more serious and significant moment. For a pair-forming species this makes sense.

Inside the vagina there are two zones of unusual sensitivity, and these sexual 'hot spots' will be discussed later. The lower third of the vagina – the part nearest its opening – is surrounded by muscular tissue. This tissue controls the size of the vaginal opening, making it tighter in younger women. In older women who have already given birth, these muscles become weaker and some of the vaginal tightness is lost. Because this tightness appeals to males, there is now a cosmetic operation available for re-tightening the vagina – creating the infamous 'designer vagina' – the genital equivalent of a face-lift. The area of the upper two-thirds of the vagina – its inner section – is less muscular and can be more easily expanded to accommodate the male penis. At the inner end of the vagina is the cervix – the mouth of the uterus.

During the sexual act, intense arousal has the effect of increasing the dimensions of the vagina, but even at its largest it is always short enough to allow the penis to reach its far end, where the sperm can be ejaculated against the cervical opening. Through this they swim on their great journey across the uterus to the Fallopian tubes where, if the timing is right, they will meet a minute egg descending, and one of them will join with it to start a new life.

Although the female's ovaries contain literally thousands of eggs, she sheds no more than 400 during her reproductive lifetime. They mature at the rate of one a month and are fertile only as they pass down the fallopian tubes, a 10-cm (4-inch) journey which takes them several days.

In addition to the vaginal passage and its surrounding labia, the female genitals also boast four sexual 'Hot Spots'. These are small zones of heightened erotic sensitivity, the stimulation of which during the mating act helps to bring the female nearer to an orgasmic condition. They are: the Clitoris, the U-spot, the G-spot, and the A-spot. The first two are outside the vagina, the second two inside it:

The Clitoris. This is the best known of the female genital hot spots, located at the top of the vulva, where the inner labia join at their upper ends. The visible part is the small, nipple-sized, female equivalent of the tip of the male penis, and is partially covered by a protective hood. Essentially it is a bundle of 8000 nerve fibres, making it the most sensitive spot on the entire female body. It is purely sexual in function and becomes enlarged (longer, more swollen, more erect) and even more sensitive during copulation. During foreplay it is often stimulated directly by touch, and many women who do not easily reach orgasm purely from vaginal stimulation find it easier to climax from oral, digital, or mechanical stimulation of the clitoris.

An Australian surgeon recently reported that the clitoris is larger than previously thought, much of it being hidden beneath the surface. The part that is visible is simply its tip, the rest of its length – its shaft – lying beneath the surface and extending down to surround the vaginal opening. This means that, during pelvic thrusting, its concealed part will be massaged vigorously by the movements of the inserted penis. There will therefore always be some degree of clitoral stimulation, even when the tip is not touched directly. The clitoral shaft is, however, less sensitive than the exposed tip, so that direct contact with the tip will always have a greater impact on female arousal. Some women claim that, by employing a rhythmic, downward roll of the pelvis, they can create a direct friction on the clitoris tip while the male is making pelvic thrusts, and can in this way magnify their arousal, but this requires a more dominant role for the female, which is not always accepted by the male.

The U-Spot. This is a small patch of sensitive erectile tissue located just above and on either side of the urethral opening. It is absent just below the urethra, in the small area between the urethra and the vagina. Less well known than the clitoris, its erotic potential was only recently investigated by American clinical research workers. They found that if this region was gently caressed, with the finger, the tongue, or the tip of the penis, there was an unexpectedly powerful erotic response.

While on the subject of the female urethra, it is important to mention 'female ejaculation'. In the male, the urethral tube delivers both urine and seminal fluid containing sperm. In the female it is usually believed that it delivers only urine, but this is not the case. When there is an unusually powerful orgasm, some females may emit a liquid from their urethral openings that is not urine. There are specialized glands surrounding the urethral tube, called Skene's glands, or para-urethral glands, similar to the male's prostate, and under extreme stimulation they produce an alkaline liquid that is chemically similar to male seminal fluid. Women who experience ejaculation (which ranges in quantity from a few drops to a few tablespoonfuls), sometimes imagine that the extreme muscular exertions of their climactic moments have forced them into involuntary urination, but this is simply because they do not understand their own physiology. Nor, incidentally, did some medical authorities, who insisted that ejaculating women were suffering from 'urinary stress incontinence' and suggested operations to cure it. (One man recently sued for divorce because he believed that his wife was urinating on him, such is the ignorance of female genital activity.)

It is not clear what the value of this female ejaculation can be, as its occurrence is clearly a little late to act as an aid to lubrication. Vaginal lubrication is, in fact, carried out by the walls of the vagina themselves, which rapidly become covered in a liquid film when female sexual arousal first begins.

The G-Spot, or Grafenberg Spot. This is a small, highly sensitive area located 5–8 cm (2–3 inches) inside the vagina, on the front or upper wall. Named after its discoverer, a German gynaecologist called Ernst Grafenberg, it is sometimes romantically referred to as the Goddess Spot. Research into the nature of the female orgasm, carried out in the 1940s, led to the discovery that the female's urethral tube, that lies on top of the vagina, is surrounded by erectile tissue similar to that found in the male penis. When the female becomes sexually aroused, this tissue starts to swell. In the G-spot zone this expansion results in a small patch of the vaginal wall protruding into the vaginal canal. It is this raised patch that is, according to Grafenberg, 'a primary erotic zone, perhaps more important than the clitoris'. He explains that its significance was lost when the 'missionary position' became a dominant feature of human sexual behaviour. Other sexual positions are far more efficient at stimulating this erogenous zone and therefore at achieving vaginal orgasms.

It should be pointed out that the term 'G-spot' was not used by Grafenberg himself. As mentioned above, he called it 'an erotic zone', which is a much better description of it. Unfortunately, the modern use of 'G-spot' as a popular term has led to some misunderstanding. Some women have been led to believe, optimistically, that there is a 'sex button' that can be pressed like a starter button, at any time, to cause an erotic explosion. Disappointed, they then come to the conclusion that the whole concept of a 'G-spot' is false and that it does not exist. The truth, as already explained, is that the G-spot is a sexually sensitive patch of vaginal wall that protrudes slightly only when the glands surrounding the urethral tube have become swollen. Several leading gynaecologists denied its existence when it was first discussed at their conferences, and a major controversy arose, but later, when it was specially demonstrated for their benefit, they changed their minds. Sexual politics also entered the debate, when certain anti-male campaigners rejected out of hand the idea that vaginal orgasm could be possible. For them clitoral orgasm was politically correct and no other would do. How they have reacted to the recent marketing of 'G-spotter' attachments for vibrators is not recorded.

Astonishingly, there have been recent reports that some women have been undergoing 'G-spot enhancement'. This involves injecting collagen into the G-spot zone to enlarge it. According to one source, 'One of the latest procedures to catch on is G-spot injection. Similar substances to those injected into the lips to plump them up can now be injected into your G-spot. The idea is that this will increase its sensitivity and so give you better orgasms.' This sounds more like an urban myth than a surgical reality, but where female sexual improvements are concerned, almost anything is possible.

The A-Spot, AFE-zone or Anterior Fornix Erogenous Zone. Also referred to as the *Epicentre,* this is a patch of sensitive tissue at the inner end of the vaginal tube between the cervix and the bladder, described technically as the 'female degenerated prostate'. (In other words, it is the female equivalent of the male prostate, just as the clitoris is the female equivalent of the male penis.) Direct stimulation of this spot can produce violent orgasmic contractions. Unlike the clitoris, it is not supposed to suffer from post-orgasmic over-sensitivity.

Its existence was reported by a Malaysian physician in Kuala Lumpur as recently as the 1990s. There has been some mis-reporting about it, and its precise position has been incorrectly described by several writers. Its true location is just above the cervix, at the inner-most point of the vagina. The cervix of the uterus is the narrow part that protrudes slightly into the vagina, leaving a circular recess around itself. The front part of this recess is called the anterior fornix. Pressure on it produces rapid lubrication of the vagina, even in women who are not normally sexually responsive. It is now possible to buy a special AFE vibrator – long thin and upward curved at its end, to probe this zone.

Students of female sexual physiology claim (perhaps overenthusiastically) that if these four erotic centres are stimulated in rotation, one after the other, it is possible for a woman to enjoy many orgasms in a single night. It is pointed out, however, that it takes an extremely experienced and sensitive lover to achieve this.

It has been claimed that two out of every three women fail to reach regular orgasms from simple penetrative sex. As mentioned above, most of them find that only digital or oral stimulation of the clitoris can be guaranteed to bring them to climax. This must mean that, for them, the two 'hot spots' inside the vagina are not living up to their name. The reason for this, it seems, is monotony in sexual positioning. A group of 27 couples were asked to vary their sexual positions experimentally, employing postures that would allow greater stimulation of the two vaginal 'hot spots', and it was found that three-quarters of the females involved were then able to achieve regular vaginal orgasms.

Finally, the changes that the female genitals undergo during sexual arousal can be summarized as follows:

Phase one: the start of sexual arousal
Vaginal lubrication begins within the first minute.
The inner two-thirds of the vaginal tube begins to expand.
The cervix and the uterus are pulled upwards.
The outer labia begin to spread apart.
The inner labia begin to swell.
The clitoral tip begins to grow in size.

Phase two: full arousal
Lubrication eases off.
The inner two-thirds of the vagina is now fully expanded.
The walls of the outer third of the vaginal tube are swollen from vasocongestion.
The vaginal opening decreases in size by 30 per cent, due to swelling of vaginal walls.
The outer labia are so spread apart that the vagina is more conspicuous.
The inner labia are now at least twice as thick.
The inner labia change colour from pink to red.
The clitoris is fully erect.

Phase three: orgasmic climax

    The outer third of the vagina undergoes rhythmic muscular contractions.

    The first, most powerful contractions occur every eight-tenths of a second.

    The number of contractions per orgasm varies from 3 to 15.

    Muscular contractions occur throughout the pelvic region (and elsewhere).

    Female ejaculation from the urethra (of non-urine liquid) may occur.

The time taken for a woman to reach orgasm can be as little as 5 minutes, but the average time, based on a study of 20,000 female orgasms, proved to be about 20 minutes. Following the orgasm, the clitoris, labia, vagina and uterus all return to their normal, relaxed condition. Some women find it possible to enjoy multiple orgasms, one after another in quick succession, while others experience such an intense first climax that they do not feel the urge to repeat it for some time.

According to a UK survey carried out in 2003, 1 in 4 women always reach orgasm when they make love; 1 in 2 usually do so, 1 in 8 rarely do so, and 1 in 20 never do. Figures like these have been used in the past to argue that women are biologically less orgasmic than men. It is much more likely, however, that men and women are equally orgasmic, but that due to cultural pressures and prudish traditions, men have become inept at fully arousing their partners. The fact that, in the same survey, 60 per cent of women mentioned that they also achieve orgasm by masturbating, suggests that the inadequacy lies, not in their sex drives, but in the sexual technique of their partners.

Considering the great delicacy, complexity and sensitivity of the female genitals, one might imagine that an intelligent species like ours would treat them with care. Sadly this has not always been the case. For thousands of years, in many different cultures, they have fallen victim to an amazing variety of mutilations and restrictions. For organs that are capable of giving so much pleasure they have been given an inordinate amount of pain.

The commonest form of assault they have suffered is circumcision. This mutilation has been rare in the West, although as recently as 1937 a Texas doctor was advocating the removal of the clitoris to *cure* frigidity. This is an isolated oddity, but in many parts of Africa, the Middle East and Asia, female circumcision has been a common and widespread practice for centuries. It is a staggering fact that, far from being an ancient memory, the practice of cutting away all or part of the external genitals of young females is still going on in more than 20 countries.

The reasons given to justify the operation include the following: If a man's penis touches a woman's clitoris, he may become sick, may become impotent, or he may die. If a baby touches its mother's clitoris when it is being born it may die. Possessing a clitoris may make a mother's milk poisonous. Possessing external genitals makes women evil-smelling. It may also drive husbands to take illegal drugs when trying to match their wives' insatiable sexual urges. Removing external genitals prevents a wide range of 'women's problems', including a yellow complexion, nervousness, ugliness, neurosis and vaginal cancer. The real reason, of course, is that, by reducing women's sexual pleasure, it helps to subordinate them to their tyrannical male partners.

How is the operation carried out? In the worst cases young girls have their labia and clitoris scraped or cut away and their vaginal opening stitched up with silk, catgut or thorns, leaving only a tiny opening for urine and menstrual blood. After the operation the girl's legs are bound together to ensure that scar-tissue forms and the condition becomes permanent. Later, when they marry, these females suffer the pain of having their artificially reduced orifices broken open by their husbands. (As if this were not enough – if the husband leaves on a long trip, his wife may be sewn up again.)

This most extreme form of female genital mutilation – infibulation – is sometimes called Pharaonic circumcision. A slightly less monstrous form involves only the removal of the clitoris and the labia. And a more moderate form, sometimes referred to as Sunna circumcision (because it is claimed to have been recommended by the prophet Mohammed), requires only the tip of the clitoris and/or the clitoral hood to be cut away.

The anti-sexual nature of these operations has been clearly expressed by one of the 'specialists' who carry it out: 'First I examine them intimately. If their clitoris hangs out and arouses them sexually by rubbing against their underwear, then that's the time it should be cut.'

Every year no fewer than 2 million small girls are held down, screaming, and, without receiving any anaesthetic, are subjected to this brutal operation. The cutting instruments are crude – razor blades, knives or scissors – there is little hygiene and there are frequent deaths, but these are always hushed up. It is defended by those who support it with the words, 'Female circumcision is sacred and life without it would be meaningless.'

The scale of this outrage against women is vast. It has been estimated that there are well over 100 million women alive today who have been mutilated in this way. Some figures, country by country, are as follows: Nigeria, 33 million; Ethiopia, 24 million; Egypt, 24 million; Sudan, 10 million; Kenya, 7 million; Somalia, 4.5 million. In addition, 90 per cent of girls living in Djibouti, Eritrea and Sierra Leone, and 50 per cent in Benin, Burkina Faso, Central African Republic, Chad, Ivory Coast, Gambia, Guinea Bissau, Liberia, Mali and Togo have been genitally mutilated. And the list goes on. Although Africa appears to be the original source of the operation, it has long since spread to the Middle East, where is it practised in Bahrain, Oman, Yemen and the UAE, and to Asia, where it is common among the Moslem populations of Malaysia and Indonesia.

Even in countries where it has been officially outlawed, it continues almost unabated. In Egypt, where it was prohibited (to no avail), the ban was overturned in 1997 by a Moslem fundamentalist who brought a court case against the government and won.

When faced with this situation, the male diplomats and politicians of the United Nations and other such impotent organizations take refuge behind convenient phrases like 'showing respect for local traditions and customs'. It is little wonder that they themselves command so little respect.

Because there has been some public questioning of the ritual recently, the mutilators (who make good money out of performing the operation) have banded together and formed a society to protect themselves. They insist that circumcising the young girls is '. . . an easy way to reduce their sexual promiscuity that would normally lead to friction in the home between husbands and wives'. And they have demanded that their governments impose a fine of half a million dollars on anyone who dares to discuss the matter further in the local media. Medical authorities are, needless to say, fighting this proposal.

In Egypt, where 3,000 girls are circumcised every day, a leading Moslem theologian has issued a *fatwa* against anyone opposing it, saying they deserve to die and referring to the operation as a 'laudable practice that does honour to women'. Since only 15 per cent of the world's population follows Islam, and almost everyone outside Islam (not to mention many inside Islam) would refuse to condone the practice, this means that this man, the Sheikh of Al Azhar, has ordered the death penalty to be carried out on, at the very least, 85 per cent of the entire human race. This holy man has no authority for his statement, since there is no mention of female circumcision in the Koran, and the claim that Mohammed said 'It is allowed [but] if you cut, do not overdo it' has been described by Moslem scholars as 'non-authentic'.

The sheikh's supporters reflect his violent posturing. When a female Egyptian reporter asked awkward questions, she was told to shut up or 'I will cut your tongue out and the tongues of those who gave birth to you'. And in a further, bizarre outburst, she was also told that if she herself had had her clitoris removed, she might have had a better complexion (one of the spurious claims for female circumcision being that it 'makes a woman's face more beautiful').

Finally, it is important to make a brief mention of the recent, fashionable practice of genital piercing. This differs in two important ways from the genital mutilation that is usually called female circumcision. First, it is voluntary and is performed only on consenting adults. Second, its stated function is 'to decorate, enhance, stimulate and arouse sexual interest in the female genitals,' rather than to destroy them.

217

Why anyone would wish to have metal studs or rings inserted into small holes bored into sensitive parts of the vulva is hard for most people to understand, but for a small minority it has become an exciting new fashion in the long history of body adornment.

The main genital piercings are as follows:

The Vertical Clitoral Hood Piercing. This is the most popular of modern genital piercings. It may consist of a thin, bent bar or rod going vertically through the hood of skin that is situated just above the clitoris, with a spherical stud attached to each end. The lower of these two studs is therefore in contact with the clitoris and may stimulate it when certain body movements occur. Or it may consist of a simple metal ring inserted vertically through the hood.

The Horizontal Clitoral Hood Piercing. In this case, the hole is pierced through from one side of the clitoral hood to the other. Again, stud-bars or rings can be inserted. The effect is said to be more decorative but less stimulating.

Clitoral Piercing. This is extremely rare, for obvious reasons. The clitoris is too sensitive and, in most cases, too small for efficient piercing.

Triangle Piercing. This is a horizontal piercing at the base of the clitoral hood. Whereas the Vertical Clitoral Hood piercing can stimulate the front of the clitoris, the Triangle piercing stimulates the back of it.

Labial Piercing. The inner labia or the outer labia are pierced with pairs of rings or studs, on either side of the clitoris or the vaginal opening.

Although this new fascination with decorative mutilation of the female genitals is probably no more than a passing fad, it is an unfortunate development at a time when so much effort is being put into trying to discourage the forcible anti-sexual mutilation of millions of girls through female circumcision. If some modern women are prepared to go to the lengths of having their genitals painfully pierced in order to follow a trivial whim of fashion, then it becomes that much harder to complain about other, more serious forms of genital mangling. But although both activities may involve surgical attacks on the sensitive female vulva, it has to be kept in mind that, in one case, the assault is supposed to enhance sexual pleasure, while in the other it is meant to destroy it.

21. THE BUTTOCKS

The buttocks have quite unfairly become the 'joke' region of the female body. They make people laugh; they are a popular subject for dirty jokes. The arse (eleventh century), the bum and the tail (fourteenth century), the butt and the rump (fifteenth century), the backside (sixteenth century), the posterior and the seat (seventeenth century), the behind, the bottom and the derriere (eighteenth century), the sit-upon and the tush (nineteenth century), the buns, the fanny and the keester (twentieth century) – whatever name they are given, the buttocks are looked upon as either ridiculous or obscene. Even when they are considered as an erotic zone, because of their proximity to the genitals, they are more likely to be pinched or slapped than caressed.

One has to search diligently through the literature to find words of praise for this region of the female anatomy. In *Lady Chatterley's Lover* D. H. Lawrence does briefly wax lyrical about the 'slumberous, round stillness of the buttocks', and Rimbaud admires them as 'two outstanding arcs', while Byron admits that the female behind is 'A strange and beautiful thing to behold.'

More recent authors have remarked, rather obscurely, that 'the ass is the face of the soul of sex' and that it offers 'a buffet of delights'. Italian film director Federico Fellini commented, equally obscurely, that 'the arsy woman is a molecular epic of femininity' – a phrase that appears to have lost something in translation. Spanish artist Salvador Dali went even further, insisting that 'it is through the arse that life's greatest mysteries can be fathomed'.

These are isolated examples, however, and it is much more

common to find the buttocks treated in a comic or vulgar manner, and this negative attitude persists despite the fact that the buttocks are specially and uniquely human. They were acquired when our ancient ancestors took the truly giant step – and stood up on their hind legs. The powerful, bulging gluteal muscles expanded dramatically, enabling their bodies to remain permanently and fully erect, and it is those muscles which gave them the pair of curved hemispheres at the base of the back which today we ungratefully find so laughable.

It is easy to see how this has come about. The buttocks are not alone. Between them lurks the anus, through which must pass, day after day, all our solid waste matter and – even more notoriously – the occasional emission of gas. Furthermore, when we bend down, the genitals swing into view, also framed by the twin curves of the buttocks. So there is no escaping excretory and sexual associations.

It follows from this that to display the buttocks is interpreted either as a gross insult – a symbolic act of defecation on an enemy – or as a gross obscenity – a shameless presentation of sexual organs. In modern society the showing of a bare behind in public can produce reactions varying from embarrassed laughter to serious complaints, outrage and even prosecution. Quite recently in Switzerland the Federal Supreme Court was wrestling with the fine point as to whether a particular buttock display was 'offensive' or 'indecent'. On this subtle distinction rested the decision concerning a conviction. A Swiss woman, during an angry quarrel with a neighbour, had suddenly 'displayed her naked posterior'. As there were children present, she was arrested on a charge of public indecency and had been found guilty by a lower court. After due deliberation the Supreme Court quashed the woman's conviction and even awarded her costs. They did this because they came to the conclusion that 'the gesture was certainly insulting behaviour and punishable as such but it could not be considered indecent because no organs of procreation were involved'. Presumably, if she had bent further forward when making her defiant gesture, her conviction would have stood.

Such extreme responses to buttock displays are rare today in the West. Streakers who expose themselves at sporting events usually produce only laughter, as do 'mooners' at American colleges who stick their behinds out of dormitory windows. As a protest, nudity is not what it used to be.

The buttock display is sometimes made more abusive by the addition of the phrase 'kiss my arse'. Taken at face value this is insulting because it demands a humiliating act of subordination. But there is more to it than that. Although probably neither the insulter nor the insulted person realises it, they are engaged in a modern version of an age-old occult practice. To understand this it is best to return first to ancient Greece.

The present-day view of the buttocks as a joke region of the body was not shared by early Greeks. To them it was an unusually beautiful part of the anatomy, partly because of its pleasing curvature but also because it made a powerful contrast with the animal rump of apes and monkeys. The human hemispheres were so different from the tough patches of hardened skin (the ischial callosities) on the lean-bottomed ape, that the Greeks saw them, quite correctly, as supremely human and non-bestial. The curvaceous Goddess of Love, *Aphrodite Kallipygos* – literally the 'Goddess with Beautiful Buttocks' – was said to have a behind more aesthetically pleasing than any other part of her anatomy. It was so revered that a temple was built in its honour – thereby making the buttocks the only part of the human body so honoured.

This early view of the buttocks as exquisitely human gave rise to a further notion. It was argued that if rounded buttocks were the hallmark distinguishing human beings from the beasts, then the monsters of darkness must lack this particular anatomical feature. So it was that the Devil gained the lasting reputation of being buttockless. Early Europeans were quite convinced that the Devil, even though he could assume human form, could never complete the transformation because, try as he might, he could never manage to simulate the rounded human buttocks. This, the most gloriously, most exclusively human feature of the body, was beyond even his fiendish powers.

This weakness was thought to be a source of great anguish to the Devil, and it provided a golden opportunity for tormenting him. To inflame his envy all that was needed was to show him your bare buttocks. Because it reminded him of his deficiency, the sudden display would force him to look away, averting his evil gaze. This protected the buttock-displaying human from the much feared 'Evil Eye' and became widely employed as a valuable device for repelling the forces of wickedness.

Buttock displays used in this special way were not regarded as vulgar or dissolute. Early fortifications and churches often displayed carvings of human females showing off their rounded buttocks to drive away evil spirits, the exposed behinds always pointing outwards from the main entrances. In the Germany of those times, if there was a particularly terrifying storm at night, women would protrude their naked buttocks from their front doors in the hope of warding off the powers of evil and avoiding a stormy death.

It is quite probable that this is how all buttock displays began and that today's streakers and mooners are carrying on an ancient Christian tradition without realizing it. With the Devil out of fashion as the great enemy, the display is now seen merely as 'rude'. From an act of religious defiance it has slipped easily into an obscene exposure of a taboo body zone.

But how does this explain the phrase 'kiss my arse'? To understand this it is necessary to examine early engravings depicting the Devil. If he does not have buttocks, then what *does* he have on his hindquarters? The answer is that where his buttocks should be he has another face. This second face is the one that was supposed to be kissed by witches as part of the ritual of the Sabbath. Accused of the filthy action of kissing the Devil's rump, they reputedly defended themselves by insisting that they had only kissed the mouth of his second face.

All these activities were, of course, the inventions of fertile medieval imaginations, but that is beside the point. The legends and beliefs handed down from one superstitious generation to the next, made it clear that 'arse-kissing' was the foul act of a follower of Satan and, as such, was an abhorrence. When the superstitions began to fade and die away, the connections were lost but, as so often happens, the popular phrase survived to be incorporated in the modern insult.

So far, the display of the buttocks has been examined purely as a hostile act – as ancient defiance or as modern insult. But there is another side to it. In a completely different context, the buttocks also transmit powerful signals of sexual appeal.

The females of many species of monkeys and apes have brightly coloured rumps. Their hind quarters become increasingly conspicuous and swollen as the time of ovulation approaches, then recede again as it passes. This means that a male can tell at a glance whether a female is sexually active. Matings usually take place only when the females are displaying their most exaggerated sexual swellings.

Human females are different. Their rumps do not rise and fall with their menstrual cycles. Their buttocks remain protuberant throughout. Matching this, sexuality also remains high. As part of her pair-bonding system, the human female has extended her sexiness so that she is always potentially responsive to the male. She will mate even at times when she cannot possibly conceive, because the function of human mating is no longer purely procreative. As a reward system it helps to cement the bond of attachment between male and female, keeping the vital family breeding unit together. As already noted, for humans, copulation is *literally* lovemaking, and it is important for the female's body to be able to transmit its erotic signals at all times.

It could be argued that if the gluteal muscles of the human rump are essentially concerned with the mechanics of standing erect, females cannot help but display permanently protruding buttocks. In their sexuality, however, the female buttocks go beyond the demands of simple mechanics. Relative to body size they are larger than those of the male, not because they are more muscled, but because they incorporate much more fatty tissue. This additional fat has been described as an emergency food store – rather like a camel's hump – but whether this is true or not, the fact that it is gender-linked automatically makes it a female sex signal.

This signal is accentuated by two other female properties: the backward rotation of the pelvis and the sway of the hips in walking. As previously mentioned, the typical female – not to be confused with the female athlete, whose body has been severely masculinized by special training – has a more arched back than the male. In a normal position of rest her behind sticks out backwards more than the male counterpart, regardless of its size. When she walks, the different leg and hip design of the female skeleton produces a greater undulation in the buttock region. To put it bluntly, she wiggles as she walks.

When these three qualities – more fat, more protrusion and more undulation – are combined, the result is a powerful erotic signal to the male. This is not because the female is deliberately thrusting out her behind and consciously wagging it at admiring males but simply because this is the way her body is designed. She can, of course, exaggerate her natural signals if she is prepared to risk caricature, and wiggle her bottom outrageously. (A zealous observer recently reported that, during an evening's performance, singer Kylie Minogue wiggled her bottom a total of 251 times.) But even if a female does nothing at all her basic anatomy will be constantly transmitting signals on behalf of her gender.

Today we may be seeing less of this female buttock signalling than was once the norm. It seems likely that the females of our early ancestors were, in fact, much bigger-buttocked than their modern counterparts. Evidence of this cannot, of course, be found in ancient skeletons, but when we look at Stone Age paintings and sculptures, huge buttocks are everywhere. They persist after the Stone Age in the prehistoric art of many cultures, but then gradually begin to disappear, dwindling to modern proportions which, although still relatively much larger than those of the males, are considerably less extreme. These early 'super-buttocks' have given rise to a great deal of speculation.

One possible scenario runs as follows: Our primeval ancestors mated from behind, like other primates, so the pre-human sexual signals of the female came from the rear, as with other species. Then, as we evolved into the erect posture and our rump muscles bulged out into buttocks, the swollen shape became the new human sex signal. Females with larger swellings on their rumps sent out stronger sex signals, and this condition then started to increase until the buttocks became huge. The sexiest females had the advantage of supernormal buttock-signals with their new super-buttocks, but these became so big that they actually began to interfere with the sexual act they were promoting. So the males solved the problem by switching to frontal copulation. As part of this new frontal approach, the breasts became permanently swollen as mimics of the large hemispherical buttocks. Now these super-breasts could also send out sexual signals, sharing the load, so to speak, with the buttocks, which could now decrease in size. This later version of the human female, better balanced and more agile, was at a considerable advantage over the fat-laden earlier model, which was gradually replaced.

If this speculative sequence is correct, we would expect to find some remnants of evidence to support it. Those remnants are to be found today in the southwestern deserts of Africa, where Bushman females still display the super-buttocks depicted in the Stone Age figures. Their remarkable contours reach astonishing proportions in some individuals and may well be showing us today what *all* our ancestral females used to look like, many thousands of years ago.

It has been argued that to compare Stone Age Europeans – who were presumably the models for the Stone Age figurines – with modern-day Bushman people living in the far south of Africa is nonsensical, but this objection overlooks the true history of the Bushman. The present-day tribespeople are not living in their remote desert because it is their favoured environment. They are there because it is the last corner of Earth where they have been able to cling on, as a disappearing branch of the human family. Their ancestors owned much of Africa and left their beautiful rock paintings behind to prove it. But they represented the Old Stone Age, the period epitomized by hunting and gathering as a way of life. With the arrival of New Stone Age peoples – the early farmers – they were driven from nearly all their territories, and today only some 50,000 survive, scarcely enough to populate a small city. In the past, however, they were one of the dominant forms of our species, and there is no reason to suppose that very large buttocks (the condition called *steatopygia)* was some kind of obscure desert rarity. It is more than likely that in the primeval hunting phase of human prehistory huge buttocks were normal for women and that the Stone Age artists based their figurines on reality rather than erotic fantasy.

When the slender, more agile females came to dominate the scene, the old big-buttock image did not fade away completely from the human unconscious mind. It still re-surfaces from time to time in rather unexpected ways. Many trivial costumes and dance movements exaggerate the buttock region. Even in staid Victorian times, with the introduction of the bustle, the male gaze was offered a new, artificial version of *steatopygia*. Hoops, padding, wire netting and steel springs were brought into play to recreate the long-lost fatty protrusions in the rump region. The elegant ladies who wore their bustles in polite Victorian society would no doubt have been horrified at such a view of their costumes, but today the comparison seems inescapable. In the twentieth century the major device for exaggerating the female buttocks was the high-heeled shoe. This type of footwear distorted female walking in such a way that the buttocks were thrust upwards and outwards more than normally and were forced to undulate even more when in motion.

Even without undue exaggeration the buttocks continue to provide one of the main erotic focuses of the modern female body. Long dresses that hide the legs are often cut so as to display the contours of the behind and clearly delineate their movements. Short garments like the 1960s miniskirt displayed the buttocks more directly and tight trousers, while obscuring the actual flesh, leave no doubt about the precise shape of the hemispheres.

In the early 1980s, there was a period of great emphasis on carefully designed, tight-fitting, high-priced jeans, deliberately intended as the perfect 'casing' for showing off this region of the body as a bold sexual signal from the newly liberated female. The author of a book called *Rear View,* published at the time and devoted exclusively to the erotic impact of female buttocks, hailed the new fashion era with these breathless words: 'The Butt Blitz began in 1979 when a spokesperson thrust her vibrating, gyrating, designerized derriere into the startled face of network television . . . It was the start of the cultural phenomenon known as Designer Jeans.'

Within a few years designer jeans were in competition with baggy, boiler suit trousers that borrowed their shape from astronaut-wear, and both styles have managed to survive alongside one another. As female trousers of one sort or another have come to dominate the world of women's fashions and skirts have increasingly lost favour among younger women, the old, crudely cut 'working jeans' have become a distant memory and more and more attention has been paid to creating female 'leg-coverings' that delineate and glamorize the buttock region.

An extreme form of this type of fashion trend appeared in 1992 when a young English designer introduced what he called 'bumsters'. These trousers had the waistband cut so low that the buttock cleft was exposed to public view. This started a phase of buttock emphasis that gave rise to such terms as 'butt couture', 'cheek chic' and 'derriere décolletage'. Not everyone in the fashion world was amused, however, one critic remarking that high fashion had sunk to the level of the 'builder's bum'.

Despite these reservations, the female buttocks were about to enjoy a new phase of erotic praise instead of the usual ridicule and, as the twentieth century came to an end, more and more young people paid attention to this region of the body. One trendy commentator was even moved to make the remark that 'bottoms are the new breasts'.

In the United States, a whole style of music, called Booty Rap, became popular. A branch of black Southern Rap music, it spread from its roots in Miami with racy song titles such as 'Free your mind and your ass will follow'.

The word 'booty' was yet another euphemism for buttocks. It originated in the twentieth century, but was then restricted to black American slang. It did not become mainstream until the beginning of the twenty-first century, when it first appeared in a general dictionary in 2002, along with its adjective 'bootylicious', which was defined as 'sexually attractive especially with curvaceous buttocks'.

The singer/actress Jennifer Lopez became the focus of this new buttock-attention in 1999 when newspapers in both Europe and America announced that she had insured her greatly admired behind for 1 billion dollars. Although she issued a denial, the fact that such a story could be invented and could make headlines was an indication of the degree of interest that existed in this part of the female anatomy as the twentieth century came to a close and the twenty-first began. In Brazil they even invented a new word to describe a woman with big, beautiful buttocks – *popozuda* – and the Brazilian music scene saw a growing cult of Popozuda Rock 'n' Roll. The skinny models, with their so-called 'heroin chic' and buttocks that were only three-quarters the size of those of other women, were suddenly out of favour.

In the UK, an annual event in which an award was made for the female who was voted 'The Rear of the Year' became increasingly popular. It had started slowly in the 1980s, but was given much more publicity as the twenty-first century dawned. On both sides of the Atlantic, demands for cosmetic improvements to the buttocks were growing. Bottom-lifters and bum-boosters were already being built in to certain female garments, but now surgery beckoned, with cosmetic surgeons reporting a surge in requests for more voluptuous bottoms, both by fat injections and silicone implants. This form of surgery costs as much as $10,000, but the high cost appears to have been no deterrent.

In addition to buttock enlargement, there has also been a great demand for buttock firming and tightening, to create a younger look as well as a more voluptuous one – a doubly improved behind. One of the world centres for this type of surgery is Brazil where there are now estimated to be no fewer than 1,600 cosmetic surgeons operating. Apparently, this type of surgery is such a dominant theme there that, when you check in to a hotel bedroom in Rio, you are likely to find leaflets advertising rival cosmetic surgeons placed alongside the inevitable Gideon Bible.

How long this fashion for firmly rounded, generously proportioned buttocks will last is hard to say, but clearly the world of fashion and popular culture keeps on returning to the primeval buttock region as an erotic focus. We may have long since given up locomotion on all fours but the sexual rump of the female refuses to fade from the unconscious mind of the male. It has even been suggested that the universal symbol of love, the stylized heart shape, is in reality based on the buttocks. It certainly looks very little like the real heart, but with the cleft in its upper surface it does have an uncanny resemblance to female buttocks seen from behind. Here again a primeval human image may be at work.

Up to this point we have looked at insulting buttocks and sexual buttocks, but there is a third way in which this part of the body has been displayed, and that is submissively. The presentation of the buttocks in a humble bent-over posture has had an enduring role as an appeasement gesture. In this respect there is no difference whatever between the behaviour of the submissive human individual and a submissive monkey or ape. In all cases, the 'presenter' is saying 'I offer myself in the passive female role. Please show your dominance by mounting me instead of attacking me.' Subordinate monkeys of either sex will make a rump presentation to dominant monkeys, also of either sex. The dominant individuals rarely attack such a subordinate, either ignoring it, or else mounting it briefly and making a few formalized pelvic thrusts. As an appeasement display the action is valuable because it enables a weak subordinate to remain close to a powerful dominant without being attacked.

In some tribal societies it has been observed that the bow performed as a greeting ceremony is done facing *away* from the greeted person. This looks so much like a 'rump presentation' that it is hard not to see it as related to the typical primate appeasement action. A much more commonplace form of rump presentation is seen when a child is spanked as a punishment. The victim must first bend over in the primate appeasement posture and then, having adopted the very position which, were it a monkey, would save it from attack, is most unfairly assaulted with a hand, cane or whip. For some dominant humans, it seems, a humiliating rump posture is not enough.

Because of their sexual implications, interpersonal buttock contacts are also somewhat restricted. Outside the sphere of loving couples, the pat or mild slap on the behind can only be used safely as signal of friendship when there is no danger of sexual implication. Employed between friends at an ordinary social gathering it could easily be misconstrued and the pat-on-the-back is preferred to the pat-on-the-behind unless sexual innuendos are deliberately intended. The pat-on-the-behind is therefore restricted to such contexts as a parent with a very small child; or sportsmen during violent team contests. In both these cases, sexual thoughts are so remote from the relationships that no misunderstanding can arise. By contrast, elderly relatives or 'friends-of-the-family' who exploit age differences by patting the bottoms of teenage daughters –

enjoying mild sexual touching disguised as harmless pseudo-parental contact – can be the source of much annoyance.

Between lovers, buttock-clasping is common in both courtship and copulation itself. It is a frequent accompaniment of advanced stages of kissing and embracing, the back embrace being lowered to a buttock embrace as arousal increases. In old-fashioned ball-room dancing, where strangers are permitted to enjoy a frontal embrace as they dance, a male partner may exploit the situation by letting his hand shift down his partner's back towards her buttocks. In the classic film caricature of this strategy he quickly finds the offending hand returned to its original position.

During the advanced stages of copulation itself, buttock-clasping often becomes quite powerful buttock-grasping as an accompaniment to vigorous pelvic thrusting. It is during this phase of body contact that the hemispherical shape of the buttocks becomes so intimately linked in the minds of the lovers with intense sexual feelings.

It is this sexual linkage, again, that causes the occasional furore over the once notorious Italian pastime of public bottom-pinching. Any attractive young woman walking the streets of an Italian city was liable to have her buttocks pinched by admiring strangers. According to her social background she may respond with pride, amusement, irritation or outrage. The author of a satirical work entitled *How to be an Italian* lists the following 'three fundamental pinches':

The Pizzicato a quick tweaking pinch performed with the thumb and middle finger. Recommended for beginners

The Vivace a more vigorous, multi-fingered pinch performed several times in quick succession

The Sostenuto a prolonged, rather heavy-handed rotating pinch for use on 'living girdles'.

Modern feminists have long since ceased to find this subject a source of humour and have even on occasion struck back by taking to the streets in search of male buttocks for concerted pinching assaults.

As potential areas for body decoration the buttocks provide little scope. They are too private for displaying the handiwork and too sat-upon for the attachment of ornaments. Tattooed buttocks are not common, except among the true fanatics. The only example of ornamented buttocks comes from the seventeenth-century work *Man Transformed* by John Bulwer, in which he shows a particularly miserable looking native with jewels hanging from the left buttock. Bulwer comments: 'Among other filthy-fine devices of some nations, I remember . . . a certain people, who in an absurd kind of bravery, bore holes in their buttocks, wherein they hang precious stones. Which by their leaves must needs prove but an inconvenient and uneasy fashion, and very prejudicial to a sedentary life.'

Finally, there is the question of the female anus being used as a sexual orifice. It has been estimated that about 50 per cent of Western women have experimented with anal intercourse at some time in their lives. Only 1 in 10 of them find it sufficiently rewarding to make it a regular feature of their sexual activities. In some parts of the world the figure is much higher than this. A survey of 5000 households in Brazil revealed that 40 per cent of rural couples and 50 per cent of urban couples 'considered anal intercourse a normal part of sexuality'.

Anatomically, the anus is rich in nerve endings, so that it has the potential to be a source of physical pleasure. However, functionally it is an exit, not an entrance, and evolution has not designed it to welcome penetration. Biologically speaking, anal sex is therefore not a 'natural' activity and it is not aided by automatic lubrication from specialized glands or any of the other changes that come to the aid of vaginal penetration. Despite this, over the course of history, the anus has often been coerced into playing the role of a symbolic vagina. There appear to be four reasons for this:

In early centuries, before condoms were available, anal sex was used as a primitive, but efficient, form of birth control. This is explicitly shown in Pre-Columbian pottery figurines from ancient Peru, for example. Wherever a couple are shown together, having sex, vaginal penetration is shown unless there is a baby sleeping next to them. When there is a baby present – the artist's way of showing that they already have a family – the male's penetration is clearly anal.

This form of contraception has survived right up to the present day in many parts of the world, especially in Latin America, parts of Africa and the East. Wherever condoms are not available, for whatever reason – poverty, ignorance or religious dogma – there is the likelihood that, despite health risks, anal penetration will be employed as a simple form of birth control.

A second reason is that it permits young couples to engage in sexual encounters before marriage without the female partner losing her virginity. This is particularly true in certain Mediterranean cultures, where the display of bloodstained sheets on the morning following the wedding is still demanded as proof of the bride's intact hymen.

A third reason has to do with the widespread male dislike of menstrual blood. Because the human female remains sexually receptive even when she is menstruating, males often wish to enjoy sex at those times, but are inhibited from doing so because of the bleeding that is taking place. Anal sex provides them with a solution to this problem.

Finally, in addition to avoiding pregnancy, injury to the hymen before marriage, or contact with menstrual blood, anal sex is also employed as an erotic variant between couples seeking sexual novelty. Together, these reasons explain the widespread occurrence of an activity that has often been an intensely taboo subject.

22. The Legs

The erotic value of legs has long been recognized. When the inno-cent, fifteen-year-old Austrian princess, Mariana, was about to be married to Philip IV of Spain, among the wedding gifts offered to her was a pair of stockings. These were brusquely refused by an envoy with the scathing comment that 'the Queen of Spain has no legs'. When she heard of this, the little princess burst into tears, horrified at the thought that, once she was married, she would have to have her legs cut off. In reality, of course, all that the envoy had intended to convey was that, since a Queen's legs could never be seen, there was no point in adorning them with decorative stock-ings. In those days, 'show a leg' may have meant 'look lively' to a sailor, but to a high-status female is was tantamount to a sexual invitation.

What is it about female legs that has led to them being considered so sexy? Their primary function is standing and walking. They have clearly evolved as locomotory structures, and yet men everywhere are obsessed with them as erotic features of the female body. A traditional all-male locker-room question is: 'Are you a leg man or a breast man?' The term 'leg man' has become so embedded in male thinking that there is even an entire publication devoted to the interests of leg-fixated males, an advertisement proudly proclaiming that 'If you are a leg man, *Leg World* is the magazine for you.'

For some men, the obsession with female legs reaches the level of becoming a full-blown fetish. Technically this is known as 'partialism', meaning that just one part of the female body is enough to provide sexual satisfaction. An extreme 'leg man' is not interested in the rest of the female body and can obtain gratification from, for example, caressing a pair of nylon stockings.

Such behaviour is comparatively rare, but even among ordinary heterosexual males, who are still sexually interested in all parts of the female body, there does seem to be an unexplained bias in favour of the leg region. So, before examining the female legs as a walking device, it is worth investigating the reasons for their sex appeal.

The first, and perhaps most obvious, sexual connection lies in the way they are joined to one another. Every time a woman moves her legs, opens them, closes them, or crosses them tightly, she inevitably draws attention to the point where they meet – which is, of course, the focal point of male sexual interest. It is almost as if, inside the deeper recesses of the male mind, a pair of female legs acts as an arrow pointing at the sexual 'promised land' of the female crotch.

In this context, the opening wide of the female legs has always been considered an action loaded with sexual significance, even on occasions when it may be no more than a woman adopting a more comfortable resting posture. This is because face-to-face coupling favours a 'legs apart' female position and male humour has often equated female sexual 'generosity' with this posture (for example, of a sexually active female: 'she had to be buried in a Y-shaped coffin' or 'her legs were without equal – they knew no parallel').

Inevitably, etiquette books have instructed young women to avoid the parted-legs posture. Amy Vanderbilt, as recently as 1972, found it necessary to inform American women that it is 'graceful to sit with the toe of one foot drawn up to the instep of the other and with the knees close together'. All legs-together postures, whether standing or sitting, have about them an air of formality, politeness, primness or subordination. The 'proper' young lady sitting decorously on her chair, left knee touching right knee, at a social gathering, displays an essential neutrality of leg posture that gives her an air of inhibited 'correctness'.

The only alternative to legs-apart or legs-together is legs-crossed. This third basic position has about it an air of informality. In the nineteenth century, women in polite society were forbidden to adopt this posture in public and even today the stuffier books of etiquette still disapprove of it. Here is Amy Vanderbilt, doyenne of modern American manners, again: 'Crossing the legs is no longer considered masculine in women, but there are good reasons to avoid it as much as possible. First, it creates unattractive bulges on the leg and thigh crossed over. Secondly, when skirts are worn short, crossed legs can be indecent or at least immodest. Thirdly, it is said to encourage varicose veins by interfering with circulation.' She goes on to caution against the dangers of crossing one's legs when applying for a job, arguing that the informality of the posture may create an impression of immodesty, or that one is too casual.

The basis of this mood difference between the prim and proper legs-together and the relaxed and casual legs-crossed is the degree which they indicate a readiness or an unreadiness to rise from the comfortable sitting posture. The legs-together position shows a deferential readiness for action. The legs-crossed position indicates that the sitter has 'settled in' and has no intention of suddenly leaping up in an attentive fashion.

Looking more closely at the action of crossing one leg over the other, it emerges that there are nine ways in which this is done. They are as follows:

The Ankle-ankle Cross. This is the most modest and formal of the crossing postures. The amount of cross-over is very small and the position is only slightly removed from the formal legs-together.

The Calf-calf Cross. This is not a common variant. It has a similar mood-flavour to the ankle-ankle cross – formal and 'correct'. These first two versions of the leg-cross are the only ones displayed by certain high-status individuals on public occasions. The Queen of England, for instance, has never been photographed with her legs crossed above the calf.

The Knee-knee Cross. This is the first of the truly informal postures and is the most common one seen on ordinary social occasions. For females wearing skirts this is one of the actions that may lead to unintended thigh exposure. It is therefore available for (conscious or unconscious) sexual displays.

The Thigh-thigh Cross. This is a more extreme version of the last one, in which the legs are crossed as tightly over one another as possible. Because of the design of the (wider) female pelvic girdle, this particular posture is easily adopted by women but is rarely performed by men.

The Calf-knee Cross. The Ankle-knee Cross; and The Ankle-thigh Cross. These three related postures involve the pulling up of one leg high on the other. It is a form of leg-crossing which, if performed by a female wearing a skirt, would expose not only her thighs but even the region of her crotch. It is therefore almost entirely limited to males and to the occasional woman wearing trousers. Because of its masculine bias it is favoured by macho males who wish to emphasize their gender (or by females who wish to show that they are 'one of the boys').

The Leg Twine. In this form of crossing, one leg is twisted around the other and held there by the entwined foot. This position transmits a powerful feminine signal, because most males find it impossible to perform. Again, it is the wider female pelvis that is responsible for this difference.

The Touching-foot Cross. In this special type of leg crossing, the crossed-over foot comes to rest alongside the calf of the other leg. This is another predominantly female posture, the action involved being extremely uncomfortable for a male, again because of his pelvic design.

These forms of leg-crossing appear repeatedly at almost every formal social gathering and represent a form of body language that transmits subliminal mood signals from person to person. Apart from the gender signals already mentioned, they can be used to signal like-mindedness between two female friends. If two women think alike on a particular issue, then they are highly likely to adopt similar forms of leg-crossing as they sit together talking. If one is much more dominant than the other, however, and is asserting her status, she will almost certainly adopt a different type of leg-crossing from her subordinate. Her legs transmit the unspoken message: 'I am different from you.'

When women are sitting side-by-side, the direction of their leg-cross is also significant. If they are friendly they point the top leg towards their companion. If they are unfriendly this leg points away and helps to bias the body in that negative direction.

There is one final, significant element in leg-crossing, and this has to do with how tightly the crossed-over legs are clamped together. In general, it is safe to say that the tighter the cross, the more defensive the mood of the woman concerned. The legs-apart posture discussed earlier revealed a basic confidence in the performer. In a sense, legs-crossed is the opposite of legs-apart and it has been suggested that, because of this, all people with crossed legs are defensive. This an oversimplification, because many people feel more comfortable in a crossed posture and adopt it even when they are alone. But it is true to say that when someone feels ill-at-ease in company they are likely to clamp their legs together rather more forcefully than when they are completely relaxed, and this element of their posture does not go unnoticed, even if their companions are unconscious of their reactions. The leg-twinings and thigh-crossings are the variants that display this type of crotch-defence most clearly.

If a woman over-does this crotch-defence action and starts to clamp her legs together in an extreme way, or wraps them almost painfully tightly around one another, this ceases to be protective and begins to develop a special kind of sexual flavour because 'the lady protests too much'. In fact, so strong are the sexual signals transmitted by female legs that only a relaxed intermediate between the two extremes of tight shut and wide open can be used without drawing sexual attention to them.

Another sexual aspect of the legs is the way that they have been concealed by clothing. Throughout history the major religions have preferred to see female legs completely covered up – yet another admission of their erotic potential. Where women have fought against this, they have shortened their skirts more and more. Each

238

degree of leg exposure has been hailed as wantonly licentious by puritanical authorities, but then often goes on to become the accepted norm. To shock, the exposure then has to go further still, until the entire leg is visible to the naked eye, with only the crotch area covered by a narrow band of cloth.

At different periods in Western culture the amount of female leg flesh visible to male eyes has varied considerably. In the last century female legs disappeared altogether for long periods and even a brief glimpse of an ankle was considered shocking. So intense and complete was this suppression of the 'erotic leg' that even the word itself became prohibited in polite circles. In the United States legs were called 'limbs'. Other euphemisms for legs included 'extremities', 'understandings', 'underpinners' and 'benders'. At table, a chicken leg became 'dark meat'.

Today it is hard for us to comprehend a social climate in which such extremes of prudery could flourish, but the fact remains that legs were a taboo subject for a very long time. Only after the First World War did they emerge from hiding and even then they caused many a raised eyebrow. The rebellious young females of the 1920s were boldly exposing their calves and even their knees and this was too much for some men. They insisted that the new fashion was causing a decline in moral standards and that the 'modern girl' was behaving like a harlot. There were many cases of employees being forbidden to wear the new, shorter skirts to work. One man, described as a distinguished lawyer, complained that 'The provocation of silken leg and half-naked thigh . . . was devastating and overwhelming.'

The main significance of such comments is that they reveal, yet again, the extremely potent sexual signalling that emanates from young female legs. The reason is obvious enough. The more of a pair of legs that becomes visible, the easier it is to imagine the point where they meet. It would be an error to conclude from this, however, that changes in skirt length during the twentieth century reflect nothing more than fluctuations in the sexual vigour of society. If we look at the rise and fall of the skirt, decade by decade in the twentieth century, it is clear that short skirts arrived in periods of economic buoyancy and long skirts reappeared during periods of economic decline. The short skirts of the roaring 20s were replaced by the long skirts of the depressed 30s; the long skirts of the austere postwar period in the late 40s were replaced by the tiny miniskirts of the swinging 60s. These in turn made way for the long skirts of the seventies' recession. It is as if young females, influenced by the mood of society, revealed their level of optimism and self-confidence by the level of their hems. To the extent that an optimistic attitude went with a lively sexuality it can be said that shorter skirts did reflect a society with greater sexual energy, but this is clearly only part of the story. The longer-skirted phase in the 70s, for example, was certainly not an outcome of prudishness.

The fact is that both the short skirt and the long skirt have sexual potential, with regard to the exposure of the legs. The short skirt has the advantage that it shows off the 'lower limbs' all the time, so that they are repeatedly displayed to males; but it also has the disadvantage that familiarity exhausts the male response. As any strip-tease dancer knows, you always start your act fully clothed and it is the slow removal of the skirt that makes the emergence of the legs such a strong sexual stimulus. The long skirt therefore has the advantage that it can make a powerful impact when raised or removed, but the disadvantage that for much of the time it blocks sexual signals from the legs.

What the very short skirts have symbolized, more than any sexual factor, is a sense of freedom. Females in short skirts can stride and leap and step out in the world. Those in long flowing skirts or tight tubular ones are engulfed in them and held back by them. The explosion of leggy mini- and micro-skirted girls in the 60s was the result of the new-found freedom stemming from the invention of the contraceptive pill and from the boom economy. The long legs transmitted the social message: 'We young females are on the move.'

By the time the 80s had arrived it was clear where the move had taken them – to the feminist movement and a renewed struggle for true sexual equality. With this last step came another shift. While the confused economic picture gave rise to confused fashions for skirts – some long, some medium and some short – the *avant garde* of the female population were sidestepping the issue completely by switching to leg-equality: they donned male leg-attire – jeans, slacks and trousers.

These garments, which like short skirts had caused uproar when first introduced and which led to young females being thrown out of elite gatherings, quickly became acceptable in more and more contexts. (By the twenty-first century, 84 per cent of young women walking down London streets were wearing trousers in preference to skirts.)

Just like short skirts and long skirts, tight female trousers had an advantage and a disadvantage. They revealed to the naked eye for the first time the precise shape of the region where the left leg meets the right. This gave them a strong erotic potential. But at the same time they interfered with the smooth shape of the leg, adding unaesthetic folds and crinkles to the gently curving contours. They also gave the impression of a protective coat of armour, encasing the legs and robbing them of their vulnerability to male approach. In the mind's eye of the male, to lift a skirt is easy, to remove a pair of jeans is a struggle.

If the Western world has become increasingly liberal in its attitude towards leg exposure, so that women can wear short skirts, long skirts, tight trousers or loose trousers, as they wish, without any pressure on them to conform to a rigid social ruling, other parts of the globe are still highly restrictive. Moslem countries tyrannized by strict male religious leaders still permit no female leg exposure of any kind in public. Communist China has also seen severe restrictions throughout much of the twentieth century, but this is now changing thanks to what has been called the 'marketization' of the Chinese economy. Symptomatic of this, as the twentieth century drew to a close, is the fact that female legs with sex appeal started to make an appearance on Chinese television screens.

However, although in the twenty-first century there is an air of modernization sweeping through Chinese society, it has not been allowed to spread without some resistance. As recently as 1998, for example, a group of students submitted a formal complaint demanding 'a [television] screen cleansed of commercial trash that exposes the female body to sell beautification and other products'. The authorities were concerned enough to issue a ban on inappropriately exposed female legs on television, but within weeks, the attractive legs were back selling beauty products as before. The welcome liberalization of modern China now appears to be unstoppable.

A further aspect of 'lower limb' sexuality concerns their smooth-ness. A seventeenth-century poet, musing about his loved one's legs wrote: 'Fain would I kiss my Julia's dainty leg, which is as white and hairless as an egg.' The egg-smooth finish of the skin of the female leg (sometimes perfected with a little help in the bathroom) contrasts vividly with the hairiness of the male leg and this differ-ence acts as a powerful gender signal.

The use of sheer silk or nylon stockings has often been wide-spread as a way of increasing the appearance of female leg smooth-ness. Another development in this direction has been the 'spray-on stocking' or 'air stocking'. This is obtained in canisters that produce a spray-on mist of silk powder that adheres to the legs and looks like a very sheer version of the real thing. It has the advantage that it is cooler, waterproof and never ladders. In Japan, especially, this technique has proved to be a great success. In modern Japan there are over 12 million working women who are forbidden to display bare legs by the companies that employ them, and for them the spray-on solution is ideal. It gives their legs the smoothly elegant 'clothed' look for the business place, without any of the disadvan-tages of tights or stockings.

Another gender contrast is the curvaceous shape of the female leg, when compared with the more knobbly, muscular male leg. Smoothly ascending curves appeal to the male eye, again because of the way they differ from the design of the male leg, but also because they hint at a vigorous and healthy body. Emaciated, skinny legs that have sometimes been popular in the world of high fashion, and thick, flabby legs, are both unappealing to the male eye because neither suggest a young woman at the peak of physical fitness. Curvaceous legs – not too thin, not too fat, are associated (in the primeval male mind) with a body condition that bodes well for breeding. And it has been shown that, in all human cultures, fitness for breeding is one of the key elements in feminine appeal.

Finally, there is a special advantage in having very long female legs. A curious description of a sexually attractive female that is often heard is that 'her legs reached up to her armpits'. When 1000 men were asked to name the actress with the best legs, the one voted into top place (Nicole Kidman) was well known for having especially long legs. The reason for the sex appeal of long female legs is not hard to find. Adult females possess legs that are both relatively and in absolute terms longer than those of children. There is a spurt in leg growth at puberty and longer legs therefore come to signify the arrival of sexual maturity. An unusually 'leggy' young woman therefore transmits super-female signals. In the 1940s, pin-up artists and cartoonists began to exploit this feature, sometimes increasing the legs on their drawings to as much as one-and-a-half times the length of the real legs on their live models. Clearly, if they had gone too far the legs in their drawings would have looked spidery and grotesque, but there was an optimum improvement that undeniably gave the cartoon females an added sexiness.

From that time onwards, throughout the second half of the twentieth century and into the twenty-first, real females appeared to grow longer and longer legs. In truth, of course, it was simply a matter of the longer-legged individuals being increasingly favoured by fashion houses, glamour photographers and film directors. This process continued, year after year, until today it would be impossible for a short-legged model to find work in any fashion house, or in any other glamour occupation.

To sum up, female legs are sexually exciting (1) because the point where they meet is the focus of male erotic attention, (2) because their varying postures, from wide open to tightly crossed, suggest erotic preoccupations, (3) because the degree of covering with clothing permits erotic exposure of hidden flesh, (4) because their smooth curves emphasize feminine body-shapes, and (5) because accelerated limb-growth at puberty permits longer legs to transmit signals of sexual readiness.

Leaving this question of the sex appeal of the female leg, what about its biology and anatomy? The leg accounts for half the body's height. When artists are (accurately) sketching the human form they divide it up into four roughly equal parts: from the sole to the bottom of the kneecap, from the kneecap to the pubic region, from the pubic region to the nipples and from the nipples to the top of the head.

In other words, they portray the legs as being half the height of the whole body. This is the shape of the average adult.

The teenager with the longest legs in the world – measuring 124 cm (49 inches) of her 190 cm (6 foot 3 inch) frame – has legs that are proportionally 30.5 cm (12 inches) longer than the average – indicating the degree of variation that exists in adult female leg ratios.

The skeletal foundation of the leg comprises four bones: the massive thigh bone, the longest bone in the human body, called the femur; the kneecap which protects the front of the hinge joint at the base of the femur, called the patella; the shin bone, or the tibia, which articulates with the femur; and the splint bone, or fibula, which lies alongside the tibia.

Propelled by its strong, shapely legs, the female frame has sailed over 2 metres (nearly 7 feet) up into the air and has managed a long jump of 7.5 metres (nearly 25 feet). Marathon dancing has dragged on, week after week, with the participants in a state of near exhaustion, for as long as 214 days. Such feats of strength and endurance are a remarkable testimony to the evolution of the female legs during a million years of evolution.

A great deal has been written about gait. The walking styles of different individuals and of different cultures have fascinated observers for many years. Typically, the female stride is shorter than that of the male, but personal differences are enormous and many famous women have such distinctive walking actions that they can be imitated with ease. One only has to mention the names of Mae West or Marilyn Monroe to illustrate this. At a cultural level there are huge differences between, say, Japanese women and American women. The Japanese excel at the more formal gaits, while Americans are better at the more casual kinds of locomotion.

Altogether, 36 different kinds of bipedal gait have been identi-fied for the human species – from the slow stroll, at about one step per second, to the walk, at two steps per second, to the fast sprint, at four to five steps per second – but only nine of them show any gender-bias and therefore deserve a brief mention here:

The Hobble is the gait of those whose legs cannot perform full strides with comfort. The hobbler moves forward using very short steps. This is typical of women wearing very tight skirts or cramped shoes.

The Mince is a gait in which fast but very short steps are taken. In effect, it is an exaggeration of the female walking style, the short steps of which are made even more abbreviated. It is a type of gait described as displaying 'affected preciseness'.

The Glide is an elegant version of the mince. By short, delicate movements of the feet, the body seems to glide forward as if on wheels. Once common among high-status females in parts of Europe, it is now confined largely to Japan. To create its impact it requires the wearing of a full-length skirt that hides the movements of the feet.

The Bounce is a gait typical of a teenage girl, when she walks with a springy step that bounces the body with each pace. It is a joyous gait visibly demonstrating health and optimism.

The Stride is a cool but dominant gait characterized by unusually large paces. This is typical of females imitating the forcefulness of the more powerful male gait.

The Wiggle is the erotic walk of the female who wishes to display her gender signals to the maximum. The weight is placed first on one hip and then on the other. If overemphasized, this gait quickly becomes a sexual joke. Marilyn Monroe enhanced her famous wiggle by wearing high-heeled shoes that had one heel slightly shorter than the other.

The Dart is an anxious female gait, full of short, indecisive scurrying movements, this way and that, with much flutter and birdlike change of direction.

The Prance is a playful fast walk in which unnecessary springs and small leaps are made with the feet as the woman progresses forward. It is a fast version of the bounce, with a more vigorous leg action.

The Run is of special interest because the body-design of a woman forces her to perform this action is a slightly different way from a man. This is due to the way in which the female legs are attached to the pelvic girdle. Just as this anatomical design results in females being able to cross their legs in a different way (with the leg twine), so it gives them a different running gait, with a rotation element that is lacking in the male running action. This difference has been obscured by the fact that we watch female athletes running more than other, less athletic women, and top female athletes are selected (from millions of women) because of their unusually masculine gait. Their bodies lack the usual feminine curves and prominent breasts, their fat layers are greatly reduced and their running leg actions are in a frontal plane, with all traces of typical female leg rotation eliminated. These are the females runners we see so often on our television screens, but if a less muscular, more voluptuous woman is closely observed when she is, say, running to catch a bus, the typical female leg-rotation is all too obvious. This ungainly gait suggests that the female body, in its specialization as a childbearing structure, has sacrificed some of its athletic running ability, which developed as a specialization of the (primeval hunting) male.

Some forms of female locomotion are triggered by emotional conditions while others are the result of social rules. These rules have varied from epoch to epoch, and in more formal times strict conditions were laid down concerning the manner in which a lady should walk in a public place. A century ago she was told to avoid 'athletic striding', 'happy-go-lucky ambling', trotting, shuffling and rushing. An early etiquette book describes a woman with a socially acceptable gait in the following words: 'Her body is perfectly balanced, she holds herself straight, and yet in nothing suggests a ramrod. She takes steps of medium length and walks from the hip and not the knee. On no account does she swing her arms, nor when walking does she wave her hands about in gesticulation.'

These rules of 'good deportment' sound strange today, when a woman simply walks out of the front door and down the street without giving a passing thought to the way she is placing one foot in front another. This new informality has allowed personal gaits to develop unhindered by the restrictions of etiquette, and provides a much greater variety of walking actions.

Finally, one female leg action that deserves a mention, despite the fact that it is rapidly disappearing from modern society, is the curtsey – a salutation in which one foot moves back behind the other, and then both legs are bent slightly at the knee. In origin it is a partial kneeling action – a token half-kneel. Today this leg movement is largely limited to women greeting royalty, but in earlier times, it was widely used as a polite greeting, often combined with a bow of the head. This was performed by both sexes, but then, in the seventeenth century, these two elements – the dip of the legs and the lowering of the head – became separated from one another, the curtsey becoming exclusively female and the bow becoming exclusively male. The one context is which this gender division breaks down is in the theatre, where today the actresses tend to copy the actors and offer their audiences a bow rather than a curtsey. The exception to this is when the play they have performed is a period piece – a restoration comedy, for example – and belongs to the period when the correct form of salutation was the combined curtsey and bow. Only then does one see the original combined version being given by the actresses, who attempt in this way to remain in period.

23. THE FEET

The human foot is yet another piece of anatomy that acts as a gender signal – large for male, small for female. The female foot is both shorter and narrower than that of the male. The average length for the male foot is 26.8 cm (10.5 inches) and for the female 24.4 cm (9.5 inches). More specifically, the female heel is narrower in relation to the forefoot (or the ball of the sole of the foot), than is the case in the male.

As with other parts of the body this difference in size has been endlessly exploited and exaggerated. If a small foot is a female feature, then it follows that a tiny foot is ultrafeminine, and during the course of history countless women have suffered as a result of this. Their feet have been squeezed, squashed, cramped and crushed in pursuit of small-footed beauty. But before examining these painful modifications, what of the foot itself?

We take the act of standing upright for granted, and yet it is an extremely rare act for any mammal to perform. The human foot – a masterpiece of engineering, as Leonardo da Vinci called it – makes this possible. Structurally, the foot contains 26 bones, 114 ligaments and 20 muscles, and with these it has to maintain our balance and enable us to walk, run, leap, dance and kick. It has been calculated that, for an average, active woman, the feet hit the ground more than 270 million times during a lifetime. This is a formidable task and yet we rarely give it a moment's thought. Taking their cue from our eyes, our feet effortlessly serve us and steer us through our changing environment. Almost the only time we are reminded of how wonderfully they perform is when our eyes let us down and, in the half-dark, we take one step too many when going up or down a flight of stairs. Expecting a surface that is not there, or finding one we did not expect, leads to a sudden shock and loss of balance. In those rare moments we are reminded of just how brilliant our feet are at all other times.

When we are moving about, each foot-action consists of three elements. The first is shock-absorbing, as the foot touches the ground; the second is body-supporting, as it takes our weight; and the third is propulsion, as it helps to push us forward. This triple duty is carried out every time we take a step. To do this efficiently we have, during the course of evolution, made a small sacrifice – we no longer have opposable big toes, like other primates. The big toe has been lined up alongside the smaller ones and can no longer be used to grasp objects in the way that we use our thumbs. This makes us far less acrobatic when trying to climb trees, but it is a small loss when set alongside the huge gain we have in fleet-footed running and walking.

The specialization of the human male as a cooperative hunter meant that larger feet were a distinct advantage for him. They were needed for the chase. There was no such evolutionary pressure on the female foot and it has remained smaller and more nimble. Seeking to exaggerate this feminine quality, women have, for centuries, attempted to squeeze their feet into uncomfortably tight shoes.

Three devices have been employed by shoemakers to help their clients' feet to look smaller than they really are. The first is to make shoes too tight. The second is to make them too pointed and the third is to make them high-heeled. The first squeezes the foot, the second streamlines it, and the third makes it appear shorter by raising the position of the heel. Together, these modifications to the natural condition may create a 'sexier' foot, but they also put a serious strain on it. And it is no accident that 80 per cent of foot surgery patients are women.

The whole balance of the female body is disturbed by the extreme form of 'fashion shoe', giving rise to leg pains, backache and even headache, but the deep-seated fear of the ugliness of the big-footed body drives women on. Words like 'clodhopper' and 'goosefoot' do little to help.

A woman unfortunate enough to have a pair of large, masculine feet is looked upon as decidedly odd – so odd, in fact, that the American jazz pianist Fats Waller composed a whole song about her. It offers her nothing but ridicule and downright rejection, and includes the following lines: 'Up in Harlem at a table for two, there were four of us – me, your big feet and you. From your ankles up I'd say you sure look sweet, from there down there's just too much feet. Yes, your feet's too big. Can't use you cos your feet's too big . . . Oh your pedal extremities are colossal. To me you look like a fossil . . .'

It is little wonder, then, that many women have gone to great lengths to minimize these extremities. The passion for small feet reached such intensity in earlier centuries that some ladies of fashion were known to have had their small toes amputated to give them an extra advantage when slipping their feet into ever-more-pointed footwear.

Mention of amputation inevitably brings to mind the cruel story of Cinderella. Today's Disneyfied version of it is harmless enough, but the original was bloody and savage. A prince was looking for a wife, but to satisfy his demand for femininity, she had to have very small feet. A tiny fur slipper was used to test prospective brides. Two sisters were desperate to be chosen. The elder one tried to force her foot into the slipper, but it would not fit, so her mother told her to cut off her big toe, explaining that once she was married to the prince she would never need to walk again, so there was nothing to lose. The girl chopped off her big toe and squeezed her bleeding foot into the slipper, but as she rode away with the prince, he noticed the blood oozing from the slipper and staining her stockings. He returned her to her mother, who offered him the other daughter. This time the unfortunate girl had to have her heel cut down in size to squeeze into the slipper. Again, spurts of blood gave the game away and she too was rejected. Only then did the prince find Cinderella, whose tiny foot was a perfect fit and who became the blushing bride of the princely foot-fetishist.

The bizarre premise of this story – that a high-ranking male will find a tiny-footed female acceptable regardless of her other qualities – seems to have been overlooked by modern audiences. This is because the modern version of Cinderella has converted the two sisters into *ugly* sisters, while Cinderella is always very pretty. But this is cheating. The prince made only one demand of his bride – that her foot should fit the tiny fur shoe (not glass, by the way, which was a mistranslation of *vair* for *verre*). To understand why he put such emphasis on the foot alone, it is necessary to know that this story originated in China, where footbinding of young girls had been a common practice among families of high rank for centuries. There, the smallness of a girl's foot was the all-important mark of beauty.

Chinese footbinding began in the tenth century and lasted for just over 1000 years. Amazingly, for such a barbaric custom, it was not outlawed until the early part of the twentieth century. It took the following form. When a girl was tiny she was allowed to run about freely, but before long, usually between the ages of six and eight, she was subjected to the agony of having her toes tied in to her sole. First, her feet were washed in hot water and massaged. Then, a bandage 5 cm (2 inches) wide and 305 cm (10 feet) long was wrapped over the four small toes, bending them cruelly back on themselves. It was then wound tightly around the heel, pulling the bent toes and the heel closer together. The rest of the bandage was wrapped around and around to ensure that the foot could not be forced open again into a normal position. Only the big toe escaped this punishment and was left unbound.

Girls who cried were beaten. Despite the pain, they were forced to walk on their crushed feet in order to force the feet into accepting their new, buckled shape. Every two weeks, a new pair of shoes was put on, always 0.25cm (one-tenth of an inch) shorter than before. The aim, incredibly, was to reduce the length of the foot to one-third of its normal size – to the much-prized 'Three-inch Golden Lotus'.

By the time they were adult such girls were permanently crippled, unable to walk normally and strictly limited in the physical activities they could perform. This was the social bonus of the deformity. Not only did they have super-feminine smallness of foot, but they were also literally unable to stray from their husbands. In addition, they provided a permanent display of high status, since they clearly could do no manual labour of any kind. Only with the modernization of China in the twentieth century and the sweeping away of Mandarin society was this extraordinary form of female mutilation stamped out.

One of the reasons for the appeal of the Chinese bound foot was sexual. The Golden Lotus, as the tiny foot was called by its male admirers, had erotic significance in several strange ways. The girls' lovers were said to enjoy not merely kissing their feet during sexual foreplay, but actually taking the whole foot into their mouth and sucking it avidly. The more sadistic lovers enjoyed the ease with which they could make their women scream during lovemaking simply by squeezing their crippled feet. Furthermore, by placing the two feet together, their buckled shape formed a pseudo-orifice that could be employed as a symbolic vagina. The real vagina was also said to be improved by the stilted form of walking caused by the bound feet: 'The smaller the woman's foot, the more wondrous become the folds of the vagina.'

In addition to these and other, more outlandish erotic ideas about the Golden Lotus, there was a general sexual excitement in the idea of the helplessness of the females with bound feet. With their localized form of bondage they were at the mercy of their men, and suffered at their hands for centuries.

Leaving China, the general symbolism of the foot is often sexual elsewhere, even without the bondage factor. There has been a widely held belief that unusually large feet in a man mean that he has a large penis and that unusually small feet in a woman mean that she has a small vagina. But this is no more than a simplistic extension of the biological gender difference in foot size.

The shoe has frequently been employed as a symbol of the female genitals, and this is why 'The Old Woman who Lived in a Shoe' (in other words, whose life revolved around her genitals) 'had so many children she didn't know what to do'. It also explains why shoes are tied to the back of the cars of departing honeymooners and why a romantic lover used to drink champagne out of his lady's shoe. An old French tradition demands that the bride should keep her wedding shoes and never give them away if she wants to live happily ever after with her husband. And Sicilian girls seeking a husband always slept with a shoe under their pillows. These and many other similar customs confirm the symbolic link between the shoe and sex.

Both the shoe and the foot itself figure extensively in the strange world of the sexual fetishist. For those males who have an erotic fixation on female shoes, the style of footwear involved is usually limited to the extreme forms of high-heeled stilettos. In the bizarre world of sexual fantasy, this design of shoe becomes a brutal weapon of voluntary torture for the male masochist, his dominant female partner trampling on his body with her sharply pointed heels.

The naked foot plays a different fetish role. It is kissed, caressed, licked and sucked. The obsessed male may or may not be in a subordinate role. He may be cowering at the feet of a dominant female partner, obeying her orders to attend to her feet. Or he may, in complete contrast, be in a dominant role himself, with a helpless female partner being gently tortured by having her feet sensitized by the application of his mouth, to a point that is beyond pleasure. Or there may be no sado-masochistic elements involved, the naked female foot being stroked and kissed to enhance normal foreplay arousal.

To most people, all this sexual attention to the humble foot seems decidedly odd. Feet are, after all, encased for most of the day in a leather covering that encourages the development of bacterial and even fungal growth. Foot odour is so widespread that special products are sold to combat it. This does not give the feet much of a boost in the erotica stakes. So why do certain individuals still find this anatomically non-sexual part of the body so sexually stimulating? Why did no less an expert than Casanova remark that 'men with strong sexual appetites feel a marked attraction to the female foot'?

There are two answers. One has to do with scent glands and the other with sexual symbolism. There are specialized skin glands on the feet that transmit personal signals about their owners. If we walked barefoot we would automatically leave a trail of personal fragrance wherever we went. To this day, some tribal people can detect this fragrance and by sniffing a path can tell who has walked down it and when. If this seems farfetched one only has to remember that a bloodhound can track a human scent that is 24 hours old, over a distance of 3 miles in as little as 18 minutes, ignoring other strong scents that may crisscross its path.

In our ancient, unclothed past, this scent-signalling system of the human foot undoubtedly had some use, but in modern urban life all this has changed. Inside our airless shoes, bacteria quickly breed and our scent secretions soon decay. If we do not change our footwear and wash our feet daily, their pleasant, natural fragrance quickly deteriorates and our feet begin to smell. During the stress and agitation of modern living, we sometimes notice that our palms are sweating, but we fail to detect that our shoe-encased feet are also sweating in a similar way. This moisture cannot evaporate as nature intended and our feet suffer as a result.

It is not surprising, therefore, that to many people the idea of foot-kissing or toe-sucking is repugnant rather than enticingly erotic. They are thinking of the modern foot as it all too often is, rather than as it should be. When it is freed of its shoe-prison in the bedroom, bathed and cleansed in the bathroom, and then presented for a lover's caress, it is an entirely different sexual proposition. It is suddenly the fragrant object that nature intended it to be and close contact with it can be exciting both for the foot-owner and for her attentive lover.

Beyond this primeval appeal there is also a symbolic attraction in operation. Sucking a female toe provides the amorous male with a sensation that he is closing his lips over a giant nipple, a huge clitoris, or even a female tongue. Again, to some, these symbolic equations may seem farfetched, but it is well known from many psychiatric studies that, at moments of sexual arousal, certain parts of the body easily become 'anatomical echoes' of other organs. Lips become labia, the mouth cavity becomes a vagina, stiff fingers become a penis and breasts become buttocks, inside the sexually excited brain.

Furthermore, while sexual foreplay is taking place, the female feet are not insensitive. The kissing, sucking and licking makes them highly responsive. Freed of their shoes, they demonstrate a response to touch that can become intensely erotic. During moments of orgasm, the toes can be seen to spread wide or curl tight, as if the feet are trying their best to join in the writhing of the climactic body-reaction.

To sum up, despite the appalling way they have been treated in recent times, the female feet remain a powerful erogenous zone, both for their owner and for her male partner.

Leaving the erotic aspects of the feet, in a non-sexual context the female foot has frequently been exploited as a focal point for high-status displays. These have taken several forms, including exotically expensive shoes, golden anklets, jewelled toe-rings, and time-consuming toenail decorations.

Some exceptionally high-status women have demonstrated their power and wealth by the size of their shoe collections. In recent times, Imelda Marcos, the 'Steel Butterfly' of the Philippines, was a striking example, forever travelling the world to buy new shoes. She reputedly owned more then 3000 pairs, housing her collection in five separate rooms in the Presidential Palace in Manila. After she and her husband were removed from power she was accused of 'putting the delectation of her feet' above the basic needs of her subjects. She countered that she had collected them as 'a symbol of love and thanks', and that, in any case, she had only 1,060 pairs, not 3,000. Curiously, of these 1,060 pairs, 1,220 pairs are now on show in the recently opened Philippines Footwear Museum, and it has been estimated that today Mrs Marcos has managed to create a new collection of 2,000 more pairs.

Even more extreme was the case of the Princess Eugénie, wife of Napoleon III, who refused to wear any pair of shoes more than once. Luckily she had very small feet, so that her daily cast-offs could be gathered up and despatched to orphanages for the bare-foot girls there to wear.

Perhaps the most extraordinary example of high-status female footwear went on display at Harrods in London in the spring of 2003. A pair of stiletto-heeled red shoes by designer Stuart Weitzman, boasting 642 rubies set in specially spun platinum, inspired by the magical ruby slippers worn by Dorothy in *The Wizard of Oz*, was put on sale at a price of 1 million pounds (approximately 1.5 million dollars).

Finally, it has to be admitted that the early craving for abnormally small female feet is a painful traditional that is still with us today. Modern shoe designers have once again been putting cruel demands on their fashionable female clients. Glamorous shoes have become increasingly narrow and pointed, and in 2003 it was predicted that future designs would be 20 per cent narrower and more sharply pointed than ever before. This has driven some women in the United States to request 'little toe removal' as a new form of cosmetic surgery. Podiatrists (specialist foot surgeons) have so far refused to provide this type of treatment, but some of them have agreed to the less drastic measure of shortening the second or third toe, by removing a small piece of bone. This enables the women concerned to squeeze their newly sculpted feet into more flatteringly tiny designer shoes. Cinderella lives.

References

THE EVOLUTION
Morris, Desmond. 1997. *The Human Sexes*. Network Books, London.

THE HAIR
Aurand, A. Monroe. 1938. *Little-Known Facts about the Witches in Our Hair. Curious Lore about the Uses and Abuses of Hair Throughout the World in all Ages*. Aurand Press, Harrisburg.
Berg, Charles. 1951. *The Unconscious Significance of Hair*. Allen and Unwin, London.
Cooper, Wendy. 1971. *Hair: Sex, Society, Symbolism*. Aldus Books, London.
Corson, Richard. 1965. *Fashions in Hair*. Peter Owen, London.
Macfadden, Bernarr. 1939. *Hair Culture*. Macfadden, New York.
McCracken, Grant. 1997. *Big Hair: A Journey into the Transformation of Self*. Indigo, London.
Powers, Rosemary. 1994. 'The Human Form in Palaeolithic Art'. Modern Geology'. 19, pp. 109–346.
Severn, Bill. 1971. *The Long and Short of It. Five Thousand Years of Fun and Fury Over Hair*. David McKay, New York.
Sieber, Roy. 2000. *Hair in African Art and Culture: Status, Symbol and Style*. Prestel Publishing, New York.
Trasko, Mary. 1994. *Daring Do's. A History of Extraordinary Hair*. Flammarion, Paris.
Woodforde, John. 1971. *The Strange Story of False Hair*. Routledge, London.
Yates, Paula. 1984. *Blondes. A History From Their Earliest Roots*. Putnam, New York.
Zemler, Charles De. 1939. *Once over Lightly, the Story of Man and his Hair*. Author, New York.

THE BROW

Cosio, Robyn, and Robin, Cynthia. 2000. *The Eyebrow*. Regan Books, New York.

Herrera, Hayden. 1983. *Frida: a Biography of Frida Kahlo*. HarperCollins, New York.

Lavater, J. C. 1789. *Essays on Physiognomy*. John Murray, London.

Parker, Nancy, and Kalish, Nancy. 2000. *Beautiful Brows: the Ultimate Guide to Styling, Shaping, and Maintaining Your Eyebrows*. Three Rivers Press, New York.

THE EARS

Mascetti, Daniela, and Triossi, Amanda. 1999. *Earrings from Antiquity to the Present*. Thames and Hudson, London.

THE EYES

Argyle, Michael, and Cook, Mark. 1976. *Gaze and Mutual Gaze*. Cambridge University Press, Cambridge.

Coss, Richard. 1965. *Mood-Provoking Visual Stimuli*. UCLA.

Eden, John. 1978. *The Eye Book*. Viking Press, New York.

Elworthy, Frederick Thomas. 1895. *The Evil Eye*. John Murray, London.

Gifford, Edward S. 1958, *The Evil Eye*. Macmillan, New York.

Hess, Eckhard H. 1975. *The Tell-Tale Eye*. Van Nostrand Reinhold, New York.

Maloney, Clarence. 1976. *The Evil Eye*. Columbia University Press, New York.

Potts, Albert M. 1982. *The World's Eye*. The University Press of Kentucky.

Walls, Gordon Lynn. 1967. *The Vertebrate Eye*. Hafner, New York.

THE NOSE

Gilman, Sander L. 1999. *Making the Body Beautiful. A Cultural History of Aesthetic Surgery*. Princeton University Press, New Jersey.

Glaser, Gabrielle. 2002. *The Nose: A Profile of Sex, Beauty and Survival*. Simon and Schuster, New York.

THE CHEEKS

Baird, John F. 1930. *Make-up*. Samuel French, New York.

Bates, Brian, and Cleese, John. 2001. *The Human Face*. BBC Books, London.

Brophy, John. 1945. *The Human Face*. Harrap, London.

Brophy, John. 1962. *The Human Face Reconsidered*. Harrap, London.

REFERENCES

Izard, Carroll E. 1971. *The Face of Emotion*. Appleton-Century-Crofts, New York.

Liggett, John. 1974. *The Human Face*. Constable, London.

McNeill, Daniel. 1998. *The Face: a Guided Tour*. Little Brown & Company, London.

Picard, Max. 1931. *The Human Face*. Cassell, London.

THE LIPS

Anon. 2000. *Lips in Art*. MQ Publications, London.

Ragas, Meg Cohen, and Kozlowski, Karen. 1978. *Read My Lips: a Cultural History of Lipstick*. Chronicle, San Francisco.

THE MOUTH

Beadnell, C. M. 1942. *The Origin of the Kiss*. Watts and Co., London.

Blue, Adrianne. 1996. *On Kissing: from the Metaphysical to the Erotic*. Weidenfeld & Nicolson, London.

Garfield, Sydney. 1972. *Teeth, Teeth, Teeth*. Arlington Books, London.

Huber, Ernst. 1931. *Evolution of Facial Musculature and Facial Expression*. Johns Hopkins Press, Baltimore.

Morris, Hugh. 1977. *The Art of Kissing*. Doubleday, London.

Perella, Nicholas James. 1969. *The Kiss, Sacred and Profane*. University of California Press, Berkeley.

Phillips, Adam. 1993. *On Kissing, Tickling and Being Bored*. Faber and Faber, London.

Tabori, Lena. 1991. *Kisses*. Virgin, London.

THE NECK

Dubin, Lois Sherr. 1995. *The History of Beads*. Thames and Hudson, London.

THE ARMS

Comfort, Alex. 1972. *The Joy of Sex*. Crown, New York.

Friedel, Ricky. 1998. *The Complete Book of Hugs*. Evans, New York.

Stoddart, Michael D. 1990. *The Scented Ape*. Cambridge University Press, Cambridge.

Watson, Lyall. 2000. *Jacobson's Organ*. Penguin Books, London.

THE HANDS

Gröning, Karl. 1999. *Hände; berühren, begreifen, formen*. Frederking & Thaler, Munich.

Harrison, Ted. 1996. *Stigmata: A Medieval Mystery in a Modern Age.* Penguin Books, New York.

Lee, Linda, and Charlton, James. 1980. *The Hand Book.* Prentice-Hall, New Jersey.

Morris, Desmond. 1997. *The Human Sexes.* Network Books, London.

Napier, John. 1980. *Hands.* Allen and Unwin, London.

Sorrell, Walter. 1968. *The Story of the Human Hand.* Bobb-Merrill Co., Indianapolis.

Ward, Anne, et al. 1981. *The Ring, from Antiquity to the Twentieth Century.* Thames and Hudson, London.

Wilson, Frank R. 1999. *The Hand.* Vintage Books, New York.

THE BREASTS

Anon. 2000. *Breasts in Art.* MQ Publications, London.

Ayalah, Daphna, and Weinstock, Isaac J. 1980. *Breasts.* Hutchinson, London.

Burr, Timothy. 1965. *Bisba.* Hercules Publishing, New Jersey.

Holledge, James. 1966. *The Cult of the Bosom. The Ups and Downs of the Bosom Over the Ages.* Horwitz, Sydney.

Latteier, Carolyn. 1998. *Breasts: The Women's Perspective on an American Obsession.* Haworth Press, Binghamton, NY.

Levy, Mervyn. 1962. *The Moons of Paradise: Reflections on the Breast in Art.* Arthur Barker, London.

Niemoeller, A. F. 1939. *The Complete Guide to Bust Culture.* Harvest House, New York.

Prose, Francine, et al. 1998. *Master Breasts.* Aperture Foundation, New York.

Snoop, Fabius Zachary. 1928. *From The Monotremes to the Madonna. A study of the breast in culture and religion.* John Bale & Co., London.

Spiegel, Maura, and Sebesta, Lithe, 2002. *The Breast Book: an Intimate and Curious History.* Workman, New York.

Stoppard, Miriam. 1996. *The Breast Book.* Penguin Books, New York.

Wilson, Robert. 1974. *Book of the Breast.* Playboy Press, Chicago.

Witkowski, G.-J. 1903. *Anecdotes historiques et religieuses sur les seins.* A. Maloine, Paris. (French text study of women's breasts in history and art.)

Yalom, Marilyn. 1997. *A History of the Breast.* Alfred A. Knopf, New York.

REFERENCES

THE WAIST

Bulwer, John. 1654. *A View of the People of the Whole World.* William Hunt, London.

Fontanel, Beatrice. 1997. *Support and Seduction.* Harry N. Abrams, New York.

Fowler, Orson S. 1846. *Intemperance and Tight Lacing.* Fowlers and Wells, New York.

Lord, William Barry. 1868. *The Corset and the Crinoline.* Ward, Lock & Tyler, London.

Moore, Doris Langley. 1949. *The Woman in Fashion.* Batsford, London.

Santé, Madame de la. 1865. *The Corset Defended.* Carler, London.

Steele, Valerie. 2001. *The Corset: a Cultural History.* Yale University Press, New Haven.

Waugh, Norah. 1954. *Corsets and Crinolines.* Batsford, London.

Zilliacus, Benedict. 1963. *The Corset.* Helsinki.

THE BELLY

Flugel, J. C. 1930. *The Psychology of Clothes.* Hogarth Press, London.

Hobin, Tina. 1982. *Belly Dancing.* Duckworth, London.

Laver, James. 1969. *Modesty in Dress.* Heinemann, London.

THE BACK

Draspa, Jenny. 1996. *Bad Backs & Painful Parts.* Whitefriars, Chester.

Inglis, Brian. 1978. *The Book of the Back.* Ebury Press, London.

THE PUBIC HAIR

Kiefer, Otto. 1934. *Sexual Life in Ancient Rome.* Routledge, London.

Licht, Hans. 1932. *Sexual Life in Ancient Greece.* Routledge, London.

Manniche, Lise. 1987. *Sexual Life in Ancient Egypt.* Kegan Paul International, London.

THE GENITALS

Bryk, Felix. 1934. *Circumcision in Man and Woman: Its History, Psychology and Ethnology.* American Ethnological Press, New York.

Chalker, Rebecca. 2000. *The Clitoral Truth. The Secret World at your Fingertips.* Seven Stories Press, New York.

Denniston, George C. and Milos, Marilyn Fayre. 1997. *Sexual Mutilations: a Human Tragedy.* Plenum Press, New York.

Dingwall, Eric John. 1931. *The Girdle of Chastity.* Routledge, London.

Ensler, Eve. 1998. *The Vagina Monologues.* Villard Books, New York.

Fisher, Seymour. 1973. *The Female Orgasm*. Allen Lane, London.
Frankfort, Ellen. 1972. *Vaginal Politics*. Quadrangle Books, New York.
Ladas, Alice Kahn, et al. 1982. *The G Spot, and other recent discoveries about human sexuality*. Holt, Rhinehart and Winston, New York.
Loughlin, Marie H. 1997. *Hymeneutics*. Associated University Presses, London.
Lowry, Thomas P. 1978. *The Classic Clitoris: Historic Contributions to Scientific Sexuality*. Nelson-Hall, Chicago.
Ridley, Constance Marjorie. 1975. *The Vulva*. Saunders, Philadelphia.
Salmansohn, Karen. 2002. *The Clitourist: A Guide to One of the Hottest Spots on Earth*. Universe Publishers, New York.
Schwartz, Kit. 1989. *The Female Member*. Robson Books, London.
Walker, Alice, and Pratibha, Parmar. 1993. *Warrior Marks: female genital mutilation and the sexual blinding of women*. Harcourt Brace, New York.
Weir, Anthony, and Jerman, James. 1986. *Images of Lust: Sexual Carvings on Medieval Churches*. Batsford, London.

THE BUTTOCKS
Aubel, Virginia (ed.). 1984. *More Rear Views*. Putnam, New York.
Hennig, Jean Luc. 1995. *The Rear View*. Souvenir Press, London.
Tosches, Nick. 1981. *Rear View*. Putnam, New York.

THE LEGS
Anon. 1970s. *Sheer Silk Legs*. J & G Trading Co., London. (Described as 'a fetish magazine for those appreciative of the sensual properties of long legs, stockings and heels'.)
Karan, Donna, et al. 1998. *The Leg*. Thames and Hudson, London.
Platinum. 1990. *Footwork*. Star Distributors, New York. (Described as 'a magazine for foot and leg worshippers'.)

THE FEET
Anon. 1989. *Foot Steps*. Holly Publications, North Hollywood, California.
Arnot, Michelle. 1982. *Foot Notes*. Sphere Books, London.
Gaines, Doug (ed.). 1995. *Kiss Foot, Lick Boot: Foot, Sox, Sneaker & Boot Worship*. Leyland Publications, San Francisco, California.
Jackson, Beverly. 1997. *Splendid Slippers: a Thousand Years of an Erotic Tradition*. Ten Speed Press, Berkeley, California.
Levy, Howard S. 1966. *Chinese Footbinding*. Neville Spearman, London.
Vanderlinden, Kathy. 2003. *Foot: a Playful Biography*. Douglas & McIntyre, New York.

REFERENCES

Wigglesworth, Linda. 1996. *The Sway of the Golden Lotus.* Chinese Costumes and Textiles, London. (Catalogue for the exhibition of a collection of Chinese shoes for bound feet.)

GENERAL

Angier, Natalie. 2000. *Woman: An Intimate Geography.* Little Brown, London.

Baron-Cohen, Simon. 2003. *The Essential Difference.* Allen Lane, London.

Biss, Hubert E. J. 1951. *Atlas of the Anatomy and Physiology of the Female Human Body.* Baillier, Tindall & Cox, London.

Boston Women's Health Collective. 1976. *Our Bodies, Ourselves: A Book By And For Women.* Simon & Schuster, New York.

Broby-Johansen, R. 1968. *Body and Clothes.* Faber, London.

Campbell, Anne (ed.). 1989. *The Opposite Sex: the Complete Guide to the Differences Between the Sexes.* Ebury, London.

Cassou, Jean, and Grigson, Geoffrey. 1953. *The Female Form in Painting.* Thames and Hudson, London

Comfort, Alex. 1967. *The Anxiety Makers.* Nelson, London.

Comfort, Alex. 1972. *The Joy of Sex.* Crown, New York.

Davis, Kathy. 1995. *Reshaping The Female Body.* Routledge, New York.

Devine, Elizabeth. 1982. *Appearances. A Complete Guide to Cosmetic Surgery.* Piatkus, Loughton.

Dickinson, Robert Latou. 1949. *Human Sex Anatomy.* Williams & Wilkins, Baltimore.

Dolezal, Seemanthini Niranjana. 2000. *Gender And Space: Femininity, Sexualization And The Female Body.* Sage Publications, New Delhi.

(Editors). 1989. *Woman's Body – An Owner's Manual.* Bantam Books, New York.

Ford, Clellan S., and Beach, Frank A. 1952 *Patterns of Sexual Behaviour.* Eyre & Spottiswoode, London.

Fryer, Peter. 1963. *Mrs Grundy. Studies in English Prudery.* Dennis Dobson, London.

Gabor, Mark. 1972. *The Pin-up, a Modest History.* Pan, London.

Gamman, Lorraine, and Makinen, Merja. 1995. *Female Fetishism.* New York University Press, New York.

Gardiner, Leslie E. 1971. *Faces, Figures and Feelings.* Burstock Courtenay Press, Brighton.

Garland, Madge. 1970. *The Changing Form of Fashion.* Dent, London.

Ghesquiere, J., et al. 1985. *Human Sexual Dimorphism.* Taylor & Francis, London.

Gifford-Jones, W. 1971. *On Being A Woman*. McMillan, New York.

Goldman, George D., and Milman, Donald S. (editors). 1969. *Modern Woman: Her Psychology & Sexuality*. Charles C. Thomas, Springfield, Illinois.

Goldstein, Laurence (ed.). 1991. *The Female Body: Figures, Styles, Speculations*. University of Michigan Press, Ann Arbor.

Goodman, W. Charisse. 1995. *The Invisible Woman: Confronting Weight Prejudice in America*. Gurze Books, Carlsbad, C.A.

Gröning, Karl. 1997. *Decorated Skin*. Thames and Hudson, London.

Guthrie, R. Dale. 1976. *Body Hot Spots*. Van Nostrand Reinhold, New York.

Hales, Dianne. 1999. *Just Like a Woman*. Virago Press, London.

Jacobus, Mary, Keller, E. F., and Shuttleworth, S. (eds.). 1990. *Body/Politics. Women and the Discourses of Science*. Routledge, London.

Jennings, Thomas. 1971. *The Female Figure in Movement*. Watson-Guptill Publications, New York.

Katchadourian, Herant A., and Lunde, Donald T. 1975. *Biological Aspects of Human Sexuality*. Holt, Rinehart & Winston, New York.

Kiefer, Otto. 1934. *Sexual Life in Ancient Rome*. Routledge, London.

Kinsey, Alfred C., et al. 1953. *Sexual Behavior in the Human Female*. Saunders, Philadelphia.

Krafft-Ebing, Richard von. 1946. *Psychopathia Sexualis*. Pioneer, New York.

Kupfermann, Jeanette. 1979. *The MsTaken Body*. Robson, London.

Lang, Theo. 1971. *The Difference Between a Man and a Woman*. Michael Joseph, London.

Lanson, Lucienne. 1975. *From Woman to Woman: A Gynaecologist Answers Questions About You and Your Body*. Alfred A. Knopf, New York.

Licht, Hans. 1932. *Sexual Life in Ancient Greece*. Routledge, London.

Lloyd, Barbara, and Archer, John. (eds.) 1976. *Exploring Sex Differences*. Academic Press, London.

Lloyd, Charles W. 1964. *Human Reproduction and Sexual Behavior*. Kimpton, London.

Maccoby, Eleanor (ed.). 1967. *The Development of Sex Differences*. Tavistock, London.

Markun, Leo. 1927. *The Mental Differences Between Men and Women: Neither of the Sexes is to an Important Extent Superior to the Other*. Haldeman-Julius Publications, Girard, Kansas.

REFERENCES

Masters, William. H., and Johnson, Virginia. E. 1966. *Human Sexual Response*. Churchill, London.

Masters, William. H., Johnson, Virginia. E., and Kolodny, Robert C. 1985. *Sex and Human Loving*. Little Brown, Boston.

McDowell, Colin. 1992. *Dressed to Kill*. Hutchinson, London.

Montagu, Ashley. 1954. *The Natural Superiority of Women*. Allen and Unwin, London.

Moore, Doris Langley. 1949. *The Woman in Fashion*. Batsford, London.

Morris, Desmond. 1967. *The Naked Ape*. Jonathan Cape, London.

Morris, Desmond. 1969. *The Human Zoo*. Jonathan Cape, London.

Morris, Desmond. 1971. *Intimate Behaviour*. Jonathan Cape, London.

Morris, Desmond. 1977. *Manwatching*. Jonathan Cape, London.

Morris, Desmond, et al. 1979. *Gestures*. Jonathan Cape, London.

Morris, Desmond. 1983. *The Book of Ages*. Jonathan Cape, London.

Morris, Desmond. 1985. *Bodywatching*. Jonathan Cape, London.

Morris, Desmond. 1994. *Bodytalk*. Jonathan Cape, London.

Morris, Desmond. 1994. *The Human Animal*. BBC Books, London.

Morris, Desmond. 1997. *The Human Sexes*. Network Books, London.

Morris, Desmond. 1999. *Body Guards*. Element Books, Shaftesbury.

Morris, Desmond. 2002. *Peoplewatching*. Vintage, London.

Morgan, Peggy (ed.). 1996. *The Female Body: An Owner's Manual*, Rodale Press, Emmaus, Pennsylvania.

Nicholson, John. 1993. *Men and Women. How Different are They?* Oxford University Press, New York.

Parker, Elizabeth. 1960. *The Seven Ages of Woman*. Johns Hopkins, Baltimore.

Ploss, Herman Heinrich, et al. 1935. *Woman. An Historical, Gynaecological and Anthropological Compendium*. Heinemann, London.

Prevention Magazine, 1996. *The Female Body*. Rodale Press, Emmaus, Pennsylvania.

Psychology Today, 1973. *The Female Experience*. Psychology Today, Del Mar, California.

Rilly, Cheryl. 1999. *Great Moments in Sex*. Three Rivers Press, New York.

Robinson, Julian. 1988. *Body Packaging: a Guide to Human Sexual Display*. Elysium, Los Angeles.

St. Paige, Edward. 1999. *Zaftig: the Case for Curves*. Darling and Co, Seattle.

Sherfey, Mary Jane. 1972. *The Nature and Evolution of Female Sexuality*. Random House, New York.

Short, R. V., and Balaban E. (eds.). 1994. *The Differences Between the Sexes.* Cambridge University Press, New York.

Shorter, Edward. 1983. *A History of Women's Bodies.* Allen Lane, London.

Steele, Valerie. 1985. *Fashion and Eroticism.* Oxford University Press, New York.

Steele, Valerie. 1996. *Fetish: Fashion, Sex and Power.* Oxford University Press, New York.

Stewart, Lea P., et al. 1986. *Communication Between the Sexes: Sex Differences and Sex-Role Stereotypes.* Gorsuch Scarisbrick, Scottsdale.

Thesander, Marianne. 1997. *The Feminine Ideal.* Reaktion Books, London.

Turner, E. S. 1954. *A History of Courting.* Michael Joseph, London.

Walker, Alexander. 1892. *Beauty in Woman Analysed and Classified.* Thomas D. Morison, Glasgow.

Warner, Marina. 2001. *Monuments and Maidens: The Allegory of the Female Form.* University of California Press, Berkeley.

Wildeblood, Joan. 1973. *The Polite World.* Davis-Poynter, London.

Woodforde, John. 1995. *The History of Vanity.* Alan Sutton, Stroud.

Wykes-Joyce, Max. 1961. *Cosmetics and Adornment; Ancient and Contemporary Usage.* Philosophical Library, New York.

INDEX

INDEX